HISTORY OF TECHNOLOGY SERIES

SERIES EDITOR: BRIAN BOWERS

GW00672072

A History of
Electric Light & Power

11 306 £3.50

BRIAN BOWERS, Ph.D., C.ENG., M.I.E.E.
Deputy Keeper, Department of Electrical Engineering, Science Museum

PETER PEREGRINUS LTD
In association with the
SCIENCE MUSEUM, LONDON

Previous volumes in this series

Volume 1   Measuring instruments — Tools of knowledge
            P.H.Sydenham
Volume 2   Early radio wave detectors
            V.J.Phillips

Published by Peter Peregrinus Ltd., Stevenage, UK and New York

© 1982 Peter Peregrinus Ltd.

British Library Cataloguing in Publication Data

Bowers, B.
    A history of electric light and power. — (IEE
    history of technology; 3)
    1. Electric utilities — Great Britain — History
    I. Title   II. Series
    338.4'762131'0941   HD9685.G72

ISBN 0-906048-71-0

Printed in England by Short Run Press Ltd., Exeter

For Faith

"WHAT WILL HE GROW TO?"

Contents

Foreword

The centenary of the first public electricity supply in the United Kingdom provides an appropriate point at which to survey the history of the development of electricity.

Dr. Bowers has done this in a thoroughly scholarly and technical way, with adequate references for those who wish to pursue particular aspects of the subject in more detail.

There can have been no development during the last century which has changed society more than the development of electricity. The brief span of one hundred years has seen the use of electricity in every home for lighting and as a means of removing the drudgery from domestic tasks; the development of electrical drives in industry to allow vastly increased productivity and improved working conditions; the development of electric traction – soon to be extended by the application of batteries to road vehicles on an increased scale; and, perhaps most important of all, the development of mass communications in the form of radio, television and telephone.

As a result, the quality and potential of life has been greatly enhanced but, in spite of these enormous developments, we are now entering the micro-processor age with its promises and challenges.

Like many others, I am proud to have been associated during my working life with the provision of electricity – a thoroughly worthwhile service, and I welcome Dr. Bowers' book which I am sure will constitute a standard work of reference.

Sir Francis Tombs
B.Sc., Ll.D., F.Eng., F.I.E.E.

Introduction

Electricity was first supplied to the public in the autumn of 1881 at Godalming, a small town in southern England. The Council had let a contract for street lighting, and the contractors offered a supply to any of the townspeople who wanted it. Few people took up the offer, and the system was closed after a few years. By then other supply undertakings had opened, in London, Brighton, and elsewhere, and the industry has been growing ever since.

The central theme of this book is the supply of electricity to the public. The first five chapters are about the discoveries and inventions in the first three-quarters of the nineteenth century which laid the technical basis for the electricity supply industry. Chapters 6, 7 and 8 are about the development of the practical generators, arc lamps and filament lamps on which the new industry was based. The growth of the supply industry and the legislative framework in which it grew are considered in Chapters 9, 10 and 11. Chapter 12 is about measurement: the science of electricity and the business of electricity supply both required trustworthy measurements. The customer's use of electricity is discussed in Chapters 13 to 16. The first electricity supply was for lighting, but lighting is mainly required when it is dark. Suppliers sought an 'off-peak' load to make use of their expensive plant during the day. The first new load was heating and cooking. Electric motors followed close behind, bringing a new source of power to homes and factories. Finally, Chapter 17 draws together the various strands which make the Electrical World of today.

Inevitably there are topics which have not been covered. The electrical achievements of the hundred years since electricity was first offered to the citizens of Godalming would fill many volumes. I hope this one volume will provide a summary of those achievements that will interest both today's electrical engineers and all who like to be informed about the world in which we live.

Many people have contributed to this book, perhaps without knowing it. Discussions with friends and colleagues in the Science Museum, and in the History of Technology Professional Group of the Institution of Electrical Engineers, have often sparked off lines of enquiry and given hints on useful sources. I am grateful to them all for sharing an interest and stimulating many thoughts. For those who would like to know more, the books and journals mentioned in the references, the Archives of the IEE, and the published volumes of the IEE's annual 'History of Technology' weekend meetings will provide a mine of information and many hours of fascinating reading.

My son, Keith, read the text and made many helpful comments.

I am grateful above all to my wife, Faith, for encouraging me to write this book and for practical help, of which typing the whole text was but a part.

Brian Bowers
April 1981

Abbreviations

Amer. J.Sci.	American Journal of Science
Electr. Rev.	Electrical Review
Electr. Times	Electrical Times
IEE	Institution of Electrical Engineers
JIEE	Journal of the Institution of Electrical Engineers
J. Soc. Tel. Eng.	Journal of the Society of Telegraph Engineers
Mech. Mag.	Mechanics' Magazine
Min. Proc. ICE	Minutes of the Proceedings of the Institution of Civil Engineers
Phil. Mag.	Philosophical Magazine
Phil. Trans.	Philosophical Transactions of the Royal Society
Proc. IEE	Proceedings of the Institution of Electrical Engineers
Proc. R. Soc.	Proceedings of the Royal Society
Trans. & Proc. of the London Electrical Society	Transactions and Proceedings of the London Electrical Society

The photographs of Faraday and his diaries and apparatus are by courtesy of the Royal Institution.

Illustrations whose source is not indicated (or apparent from the text) are Science Museum photographs.

Static and Current electricity

1.1 Electrostatics and magnetism

The word 'electricity' is derived from the Greek for amber. Thales of Miletus (640-548 BC), one of the pioneers of ancient Greek science, is said to have been the first to observe that when rubbed amber has the property of attracting light bodies.

The attractive power of lodestone, a mineral containing magnetic iron oxide, was mentioned by Lucretius (99-55 BC) and by Pliny the Elder (23-79 AD) but the of the magnetic compass for navigation began only in mediaeval times. A letter written by Peter Peregrinus in 1269 gives an account of contemporary knowledge of the compass and includes instructions for making one.[1] He knew that a compass did not point to the true North Pole.

The scientific study of electricity and magnetism began with William Gilbert (c. 1540-1603), of Colchester. Gilbert was educated at St John's College, Cambridge and then established a successful medical practice. In 1600 he became President of the Royal College of Physicians and physician to Queen Elizabeth I. His book *De Magnete,* published in London also in the year 1600, was the result of many years spare-time interest in magnetic and electrostatic phenomena. In the book Gilbert recognises that the earth itself is a magnet and he draws, for the first time, a clear distinction between electric and magnetic attractions.

The American philosopher and statesman Benjamin Franklin (1706-1790) studied electrical discharges from objects of different shapes and proposed the protection of buildings by lightning conductors. He further said that lightning conductors should be pointed. The value of lightning conductors was not universally accepted at first (did they protect a building from lightning that was going to strike it, or did they attract lightning which would not have struck the building if there had been no lightning conductor there?) There was also disagreement about the merits of pointed and ball-ended lightning conductors. King George III was persuaded to have the pointed conductors on Buckingham Palace replaced by ball-ended ones: the argument seems to have been that Franklin was a republican, therefore his scientific views must be suspect! The points were restored later.

Early electrical experiments used static electricity produced manually by rubbing together such things as glass and cloth. The first machine to generate static electricity was probably the one made in 1660 by Otto von Guericke (1602-1686), the Mayor of Magdeburg in Germany. He made a globe of sulphur which was rotated on an axle while a cloth rubbed on its surface. The globe became charged, gave off sparks and attracted light pieces of straw.

In 1675, the French astronomer Jean Picard (1620-1682) was carrying a barometer through the street at night when he noticed a glow in the top of the barometer tube as the mercury slopped about. Francis Hauksbee (c. 1666-1713), the Curator of Experiments at the Royal Society, thought that the phenomenon observed by Picard must be electrical in origin, and he devised an experiment to test his theory. He made a machine similar to von Guericke's, except that in place of the sulphur globe he used an evacuated, blown glass bulb. When rotated and rubbed, the bulb not only became charged, attracting light objects, but also glowed internally.

Stephen Gray (or Grey) (1666-1736) explained the distinction between conductors and insulators in 1720, and showed that a hemp cord suspended by silk threads could carry the electric charge a considerable distance. The Frenchman Charles François de Cisternay Dufay (1698-1739) showed that electricity appeared in two distinct forms, which he called 'vitreous' and 'resinous', and that similar charges repelled each other while dissimilar charges attracted.

The principle of storing an electric charge in the condensor, or Leyden jar, was established about 1745. Although there is more than one claim to the discovery, the credit is usually given to Andreas Cunaeus (1712-1788) working with Pieter van Musschenbroek (1692-1761) at Leyden. In one of their experiments they had a conductor inserted through the neck of a glass jar containing water. Cunaeus attempted to remove the conductor while holding the jar in his hand and received an unexpectedly powerful shock.

Another important electric charge-storage device was the electrophorus, invented in 1775 by Alessandro Volta (1745-1827), later Professor of Natural Philosophy at the University of Pavia, Italy. In the electrophorus a disc of solid resin is electrified by rubbing and used to charge a metal disc by induction. The disc is placed on the resin, its surface discharged by a brief touch, and then removed taking a charge with it. The electrophorus permitted a small charge of static electricity to be multiplied many times, but Volta's greater contribution was to come later in the field of current electricity.

It must not be thought that the story of static electricity ends with the coming of the electric current. Two important types of electrostatic machine were developed in the nineteenth century: Armstrong's hydroelectric machine and Wimshurst's influence machine.

William Armstrong (1810-1900) later Sir William and then Lord Armstrong of Cragside was a solicitor in Newcastle-upon-Tyne with an interest in science. He became one of the leading industrialists of Victorian Britain. Armstrong heard that a strange phenomenon had been observed on the Cramlington Colliery Railway near Newcastle, which had stationary steam engines to rope-haul trucks up an incline. One day in September 1840 an engine driver noticed steam leaking from the boiler. Thinking the steam pressure must be too high, he lifted the weight on the safety valve and 'felt a curious pricking sensation in the ends of his fingers'. At the time he was standing in a cloud of steam, so was not really sure what had happened. Over the following few days he experienced the phenomenon again, and mentioned it to his workmates. They all felt the same sensation and the whole affair created considerable local interest. Armstrong heard about it and, after checking for himself, wrote to Faraday about it. The steam leaking from the boiler was clearly electrified. Armstrong wanted to know how and where it became electrified. The owners wanted to know whether something unusual, and maybe dangerous, was happening inside their boiler.

Armstrong constructed a piece of experimental apparatus, consisting of two lengths of straight glass tube joined end to end by a metal stopcock. A second connected one tube to the boiler and a third was at the free end of the tubes. By measuring the potentials at each stopcock he showed that the steam became electrified only as it entered the atmosphere: it was not electrified either in the boiler or in the tubes.

In 1842 Armstrong continued his experiments by constructing an iron boiler about one metre long standing on glass legs. Steam left the boiler through a hardwood nozzle, and he found that the steam was positively charged. The following year he had an even larger machine made at the London Polytechnic Institution. It had 46 separate steam jets and could produce sparks 56 cm long. Armstrong's research on the electrification of steam and his 'hydroelectric machine' (Fig. 1.1), as it was called, brought him scientific recognition, and in 1846 he was elected a Fellow of the Royal Society.[2]

The hydroelectric machine was largely a laboratory curiosity, though when, in 1857, Wheatstone and Professor Frederick Abel were called in to advise the War Office on the electrical detonation of mines, they reported that, in some circumstances, the high voltage from an Armstrong's hydroelectric machine should be used.[3]

In practice high voltage magnetoelectric generators were soon available and it is doubtful whether Armstrong's machine was ever used. Although it has no practical value today, an understanding of the electrification of a fluid forced from a jet can be of importance. In December 1969 three large oil tankers were severely damaged by explosions. In each case it appears that tanks were being cleaned out with a high pressure water jet, and a sufficient charge accumulated to ignite residual vapours from the oil.

During the nineteenth century a number of machines were made in which electrostatic charges were multiplied by induction and accumulated. The machines could be described as mechanised versions of Volta's electrophorus and a Leyden jar. They were often called 'influence machines', and the best known and most successful was the machine devised by Wimshurst in 1883. James Wimshurst was a consulting engineer to the Board of Trade. In its simplest form his machine consists of two circular discs of thin glass revolving close together in opposite directions on a horizontal shaft. The discs are driven by belts from large pulleys turned by hand, one of the belts being crossed to obtain the difference in direction. The discs are varnished and each have a dozen or more sectors of tinfoil glued at equal intervals around their outer faces. Two metal brushes press lightly on the tinfoil as the discs revolve and metal collecting combs close to the discs collect the charges. A machine with discs 36 cm in diameter will readily give sparks 10 cm long in air.

Fig. 1.1 *Armstrong's hydroelectric machine*
The cylinder on the left is a boiler. Steam escapes through the horizontal wooden nozzle at the top, acquiring a charge by friction. A metal comb, on the top of the insulated stand on the right, collects the charge which is stored on a Leyden jar in the middle.

Wimshurst's and other influence machines were popular with physicists in the late nineteenth century because they provided a high voltage source for investigating phenomena such as high voltage discharges in gases. They were soon superseded, however, by high voltage induction coils and transformers, and relegated to the status of spectacular, but obsolete, scientific curiosities.

1.2   The electric current

In the year 1800 Volta wrote a long letter in French to Sir Joseph Banks, the President of the Royal Society of London. That letter, dated 20 March 1800 and read to the Society on 26 June 1800, contains the first description of apparatus capable of producing a continuous current of electricity.[4] Volta had studied the earlier work of another Italian, Galvani, on the electrical stimulation of frogs' legs. Luigi Galvani (1737-1798), Professor of Anatomy at the University of Bologna, had investigated the effect of discharges from frictional electric machines on dead frogs. In the course of his researches he discovered that a leg could be made to twitch with the aid of nothing more than two pieces of different metals — no electrical machine was required. Galvani concluded that the source of the phenomenon, which he called 'animal electricity' was in the nerves and muscles of the frog. Volta initially accepted Galvani's ideas, but as he continued his experiments he came to realise that the essential conditions for the production of 'animal electricity' were the presence of two different metals and a suitable liquid. The frog's leg had provided the liquid and also acted as a sensitive detector of electricity, but the leg itself was not the source of the electricity.

In his letter Volta described first his 'pile' and then his 'crown of cups' ('colonne' and 'couronne de tasses' in the original French). The letter was published, with drawings, in the Proceedings of the Royal Society.

It may be difficult for the modern reader to imagine a world without the chemical cell, or 'voltaic' cell, for producing an electric current. Before 1800 electrical phenomena were essentially transient: after Volta's discovery a continuous current of electricity could be obtained at will. Volta appreciated the significance of the discovery, and something of the excitement he felt is conveyed, even in translation, by the opening of his letter and his description of the basic pile:

I have the pleasure of sending you some striking results at which I have arrived in pursuing my experiments on the electricity produced by the simple mutual contact of different metals ... The chief result is the construction of apparatus having the properties (such as ability to give shocks) of Leyden jars which operate continuously, or whose charge is restored automatically after each discharge.

The apparatus of which I speak, and which doubtless will astonish you, is nothing but the assembly of a number of good conductors of different kinds, arranged in a certain manner, 30, 40, 60 pieces, or more, of copper or silver, each placed next to a piece of tin or, better, zinc, and an equal number of layers of water, or some other conducting liquid such as salt water, or pieces of leather or card soaked in these liquids, are placed between each pair of different metals ... Behold, this is all that makes up my novel instrument, which imitates, as I have said, the effects of a Leyden jar.

Volta described several mechanical arrangements for his 'pile', and then the 'crown of cups' which was simply a number of open glass vessels containing a suitable liquid and the metal electrodes.

Before Volta's letter was read to the Royal Society, Sir Joseph Banks showed it to Anthony Carlisle, a surgeon. Together with the chemist William Nicholson, Carlisle constructed a voltaic pile according to Volta's description and found that it could be used to decompose **water** into its elements. This discovery prompted others to make their own batteries and investigate the chemical effects of the electric current.[5]

The first mass-produced battery based on Volta's work was designed by William Cruickshank (1745-1800) of Woolwich. His battery had copper and zinc plates soldered together in pairs and then sealed with wax in a wooden trough (Fig. 1.2). Cruickshank batteries were widely used until the invention of a cell that did not polarise — the Daniell cell of 1836. Most of the basic research on electricity in the first third of the nineteenth century depended on Cruickshank batteries.

Fig. 1.2 *Cruickshank's battery*
Plates of copper and zinc soldered together are sealed with wax into grooves in a wooden trough. To activate the battery it is filled with dilute sulphuric acid.

Michael Faraday used Cruickshank batteries, and in his book *Chemical manipulation,* published in 1828, Faraday recommended that the electrolyte should consist of one volume of nitric acid and three volumes of sulphuric acid, diluted with water to one hundred volumes. This seems to be the optimum strength. Measurements made on a restored Cruickshank's battery with plates 140 x 130 mm gave an open circuit e.m.f. of 0·85 volts per cell, an internal resistance of 5·0 Ω and a maximum power output of about 10 W per square metre of electrode surface. A stronger acid than that recommended by Faraday gives a slightly greater output power, but the battery then gases freely.[6]

Although no continuous current of electricity had been available before 1800, several experimenters had shown that the discharge of electricity from a Leyden jar could fuse a fine wire and could decompose water into hydrogen and oxygen. Volta's discovery prompted many experimenters to study the properties of the electric current, especially its chemical properties. The outstanding name in this connection was Humphry Davy.

Humphry Davy (1778-1829) was born in Penzance, Cornwall; his father was a wood carver and gilder until he inherited the family farm. At the age of sixteen Humphry was apprenticed to a surgeon-apothecary, a Mr Borlase. Davy developed a great interest in chemistry, and when he was offered a post as assistant to Thomas Beddoes at the Pneumatic Institution in Bristol, Borlase agreed to release Davy from his apprenticeship. The Institution was devoted to the investigation of the physiological properties of newly discovered gases and the post suited Davy ideally. His major work there was a study of the effects of breathing nitrous oxide, or 'laughing gas', which Davy said gave him a sensation 'similar to that produced by a small dose of wine'.

Davy's work at Bristol attracted considerable attention, and early in 1801 the Managers of the Royal Institution offered him the post of 'Assistant Lecturer in Chemistry, Director of the Laboratory and Assistant Editor of the Journals of the Institution'. Davy accepted, and was given a room, coal, candles, and a salary of one hundred guineas per annum.

The Royal Institution had been founded in London in March 1799 as 'a Public Institution for diffusing and facilitating the general introduction of useful mechanical inventions and improvements and for teaching by courses of philosophical lectures and experiments the application of science to the common purposes of life'. The Institution was equipped with lecture theatre, laboratory, kitchen, and workshop, and Davy took up his appointment there in March 1801. His first task was to prepare courses of lectures on the chemistry of such practical matters as tanning, dyeing and printing, and later agriculture. Davy's lectures proved very popular, drawing large audiences from all levels of society. The poet Coleridge attended regularly and remarked 'I go to Davy's lectures to increase my stock of metaphors'.[7] Davy attracted much needed funds to the Institution, and he was then able to devote time to his own research interests, including the effects of Volta's pile on various chemical substances.

Davy's work established the science of electrochemistry, and within a few years he had isolated potassium, sodium, barium, strontium, calcium and magnesium by electrolysis of their compounds. He introduced the idea that the force binding the atoms in a chemical compound is an electric force. His current source was a Cruickshank battery. The first battery was exhausted (its zinc consumed) by 1808, and the Royal Institution organised a special subscription to purchase a battery of 2000 pairs of plates for Davy to continue his experiments (Fig. 1.3).

In the course of his electrochemical work, Davy demonstrated the brilliant light produced by an arc between two pieces of carbon connected to a high voltage supply. Although he made use of the heat of the arc, there is no evidence that he envisaged it as a practical method of illumination.[8]

Electricity and magnetism were first firmly linked together in the winter of 1819-20 when Oersted discovered that a current in a wire could deflect a compass needle. Hans Christian Oersted (1777-1851) was the son of a Danish apothecary. He became professor of physics at Copenhagen in 1806, and retained the post for the rest of his life. Oersted continued his experiments for a few months in 1820, ascertaining the precise relationship between the current in the wire and the deflection of the magnetic needle, and published his results in a Latin treatise in July 1820.[9] At that period every scientist could read Latin, and Oersted's discovery was read about and his experiments repeated throughout Europe.

Fig. 1.3 *Battery used by Davy at the Royal Institution*
Photo: Royal Institution

One man who studied Oersted's discovery was the German chemist J.S.C.Schweigger (1779-1857). He found that if the wire carrying the current was wound into a coil around the needle then the deflection was greatly increased. The arrangement became known as 'Schweigger's multiplier'. Its application as a current detector was obvious, and the name 'Galvanometer' was used by James Cumming (1777-1861), professor of chemistry at Cambridge, in 1821.[10]

Davy made one other major contribution to the progress of electrical science: he arranged for the Royal Institution to engage Michael Faraday as a laboratory assistant, and Faraday was beginning his scientific work just as Oersted's discovery came to public notice.

1.3   References

For a detailed survey of the early history of electricity, see Mottelay, Paul F.: *Bibliographical history of electricity and magnetism*, London, 1922

Heilbron, J.L.: *Electricity in the 17th and 18th centuries*, University of California Press, 1979

1   *Epistle of Petrus Peregrinus on the magnet*, mediaeval manuscript in IEE Archives and printed copy, Bernard Quaritch, 1900

2   Anderson, A.F.: 'Sparks from steam – the story of the Armstrong hydroelectric generator', *Electronics & Power*, January 1978, pp.51-53

3   Wheatstone, C., and Abel, F.A.: *Report to the Secretary of State for War on the application of electricity from different sources to the explosion of gunpowder*, HMSO, 1861

4   Volta, A.: 'On the electricity excited by the mere contact of conducting substances of different kinds', *Proc. R. Soc.*, 26 June 1800, pp.403-431 and plate XVII

5   Highton, Edward: *History and progress of the electric telegraph*, London, 1852, pp.27-29

6   The author is grateful to Mr W.A.Coates of the Royal Institution for help with the measurements described here

7   Quoted in A.Cajori: *A history of physics*, 1899, p.215

8   There are several biographies of Davy, including one by his brother, Dr. John Davy, and a brief but comprehensive booklet by Professor Ronald King of the Royal Institution:
     Davy, John: *Memoirs of the life of Sir Humphry Davy,* London 1836
     King, Ronald: *Humphry Davy,* The Royal Institution, London 1978
9   Oersted, H.C.: *Expirimenta circa effectum conflictus electrici in acum magneticum,* Copenhagen, 1820 (an English translation was published in 1877 by the Society of Telegraph Engineers: *J. Soc. Tel. Eng.,* 1877, 5, pp.459-473)
10   Cumming, J.: 'On the connextion of galvanism and magnetism', Transactions of the Cambridge Philosophical Society 1822, 1, p.269

Chapter 2

Michael Faraday

2.1 Faraday's first electrical researches

No account of the history of electrical engineering would be complete without a section on Michael Faraday (1791-1867; see Fig. 2.1). He has been called 'the father of electricity', and the highest award given by the Institution of Electrical Engineers is the Faraday Medal.

Fig. 2.1 *Michael Faraday in 1852, from a drawing by George Richmond*
Photo: Royal Institution

Faraday was born on 22 September 1791 in the Surrey village of Newington, now part of Greater London. He was the third child of James Faraday, a blacksmith, who had recently moved from Westmorland. The family soon moved again, and lived at various addresses in what is now West London. James was in poor health, and only able to work part-time. After he died in 1810 Mrs Faraday took in lodgers to support her family. Michael Faraday's formal education was minimal: in later life he said 'My education was of the most ordinary description, consisting of little more than the rudiments of reading, writing and arithmetic at a common day school. My hours out of school were passed at home and in the streets'.

In 1804 Faraday went to work as an errand boy for George Riebau, a bookseller and bookbinder. The following year, when he was fourteen, he was formally apprenticed to Riebau. Little is known about the next few years of Faraday's life, but he became a competent bookbinder and many volumes bound by him survive. The practical training in working with his hands proved invaluable later, when great manual skill was vital to his laboratory work.

With a shilling fee paid by his elder brother Robert, then a working blacksmith, Faraday joined the City Philosophical Society. There he heard lectures on a wide range of scientific subjects and met other people who shared his own increasing interest in science. In discussion after one lecture he expressed strong views contrary to the lecturer's, and as a result was invited to address the Society. Thus the greatest scientific lecturer of his time gave his first lecture. The subject was whether there were two 'electric fluids', one positive and one negative, or whether there was a single 'electric fluid', always present in some quantity. Faraday argued that there were two fluids.

Faraday's interest in science impressed a Mr. Dance, one of Riebau's customers at the shop. Dance gave Faraday tickets to hear Davy lecturing at the Royal Institution. Following the practice he had adopted at the City Philosophical Society, he took detailed notes of the lectures and bound them. Later, he sent the bound volume to Davy, requesting work in any scientific capacity. There was no vacancy at the time, but when the laboratory assistant at the Royal Institution was sacked for misbehaviour Davy offered Faraday the post. He was appointed on 1 March 1813, with a salary of one guinea a week, two rooms at the top of the Institution, and fuel and candles supplied.

From October 1813 to April 1815 Davy was travelling in Europe, and Faraday accompanied him. The tour through France, Switzerland and Italy brought Faraday into contact with many of the leading scientists of Europe, including André-Marie Ampère (1775-1836) in Paris. On his return to the Royal Institution, Faraday rapidly became established as a member of the London scientific community; he was no longer a mere laboratory assistant. Most of Faraday's work was in chemistry, but he also followed the progress of research in electricity, and knew of Oersted's discovery of the deflection of a compass needle by an electric current.

Every scientist who read Oersted's paper describing his discovery tried the experiment himself. Two main theories were produced to explain the phenomenon. Some people thought that the wire carrying a current must itself become magnetised across its width; though they could not explain why it became magnetised in that way. Others explained everything in terms of electric currents, and said that every permanent magnet must contain circulating electric currents. W.H. Wollaston (1766-1828), best known for his discovery in 1802 of the dark lines in the solar spectrum, suggested that the current took a helical path along the conducting wire, and that in consequence a wire carrying a current would tend to rotate on its own axis when a magnet was brought near. In April 1821, Wollaston and Davy tried unsuccessfully to make a wire rotate in that way.

Faraday was not actively involved in these speculations at the time, perhaps because he was busy with some research into alloy steels. However, the theories being discussed were mostly mathematical, and had little experimental basis. Faraday was no mathematician.

In the summer of 1821 he was invited to write an historical account of electro-magnetism for the *Annals of philosophy*.[1] While preparing his account Faraday repeated all the important experiments of others on the subject, and became increasingly convinced that it ought to be possible to produce continuous circular motion by using the circular magnetic force around a wire.

Success came in September 1821, when Faraday made two devices which could be called the first electric motors (Fig. 2.2). In one a bar magnet was fixed vertically in a basin and the basin nearly filled with mercury. A wire with a cork on its lower

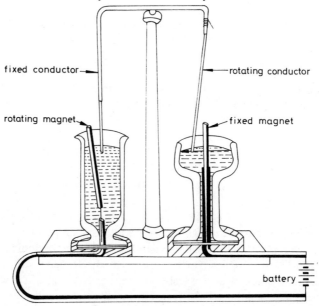

Fig. 2.2 *Diagram of Faraday's apparatus for producing rotation by electromagnetism*
On the left a magnet floating in mercury and tied at the bottom of the vessel rotates around the fixed current-carrying wire. On the right the conductor is free to move and it rotates around the fixed magnet.

end was loosely suspended from **a** point above the magnet. When a battery was connected between the suspended wire and another wire dipping into the edge of the mercury, the lower end of the suspended wire moved in circles around the magnet. In the second device one end of the magnet was tied to the bottom of the basin but the other end was free (a steel magnet floats in mercury). The loose wire of the first device was replaced by a fixed wire dipping into the centre of the mercury. When current flowed the free end of the magnet moved in circles around the fixed wire.

The experiment was repeated all over Europe. It would have assured Faraday of a place of honour in the history of electrical science, yet his greatest electrical discoveries were still ten years in the future.

During the 1820s his main scientific work was in chemistry. In conjunction with James Stodart, a surgical instrument maker, he undertook a series of experiments aimed at finding an improved steel alloy for cutting instruments. An Indian steel known as 'wootz' was favoured for these. Faraday's analysis showed that it contained small quantities of alumina and silica, and he and Stodart were able to make an alloy which closely resembled wootz. Other alloys Faraday produced were very resistant to rust, and were used in the manufacture of cutlery.

Another series of chemical researches, begun in 1825, was a study of optical glass for the Royal Society. He worked on the liquefaction of gases, and isolated benzene for the first time. The significance of benzene was only appreciated later, as organic chemistry developed. Faraday found benzene, which he called 'bicarburet of hydrogen', when analysing a troublesome by-product which accumulated in the gas cylinders of the Portable Gas Company. This company prepared a gas by heating animal oils, and it was compressed and distributed in iron cylinders for lighting. Faraday analysed a liquid residue from the cylinders which, after being purified by repeated distillation, was benzene.

In 1825 Faraday began the series of weekly evening meetings of Members of the Royal Institution, which still continue as the Friday Evening Discourses. Many of Faraday's own discoveries were first made public in Discourses that he delivered. He also began the Royal Institution's Christmas lectures for children (Fig. 2.3).

2.2 Induction

If an electric current could produce magnetic effects, could magnetism be made to produce an electric current? Faraday pondered this question from time to time during the 1820s. He was inclined to accept Ampère's idea that magnetic effects were due to tiny circulating currents in the magnet, and he reasoned that if two wires were placed side by side then a current flowing in one should produce some electric effect in the other. He tried the experiment in 1825 with two pieces of wire one and a half metres long, tied together with a single thickness of paper between them. He connected a battery to one wire, then a galvanometer to the other, but observed nothing.

Meanwhile, in Paris in 1825 Dominique François Jean Arago (1786-1853) discovered that there was a force between a moving magnet and a nearby copper plate. Arago had been puzzled by the very damped motion of a new compass, and found that it was related to the copper base of the compass. Pursuing his investigation Arago pivoted a compass needle over a copper disc and found that if the disc was turned the compass needle also rotated, or tried to, in the same direction. The effect was reduced if slots were cut in the copper disc, which suggested that the phenomenon might be associated with electric currents in the disc. It was an encouragement to pursue the search for links between electricity and magnetism.

Fig. 2.3   *Faraday giving a Christmas lecture in 1856 to an audience including the Prince Consort and the Prince of Wales*
Photo: Royal Institution

Faraday discovered electromagnetic induction on 29 August 1831, using his 'induction ring'. This was simply two coils wound on opposite sides of a soft iron ring about 2 cm thick and 15 cm in diameter. The only wire available to him was bare metal, so he insulated the turns by winding a piece of calico under each layer of wire and a piece of string between adjacent turns. The coils were in several parts so that the number of turns could be adjusted at will. Faraday drew the arrangement in his notebook (Fig. 2.4) calling the coils A and B, and noted:

> Charged a battery of 10 pr. plates 4 inches square. Made the coil on B side one coil and connected its extremities by a copper wire passing to a distance and just over a magnetic needle (3 feet from iron ring). Then connected the ends of one of the pieces on A side with battery. Immediately a sensible effect on needle. It oscillated & settled at last in original position. On *breaking* connection of A side with Battery again a disturbance of the needle.

Faraday investigated what happened when the number of turns in the coils was changed, and found that the deflection of the needle varied. He then showed that the iron ring was not essential, and that he could produce the effect with two coils

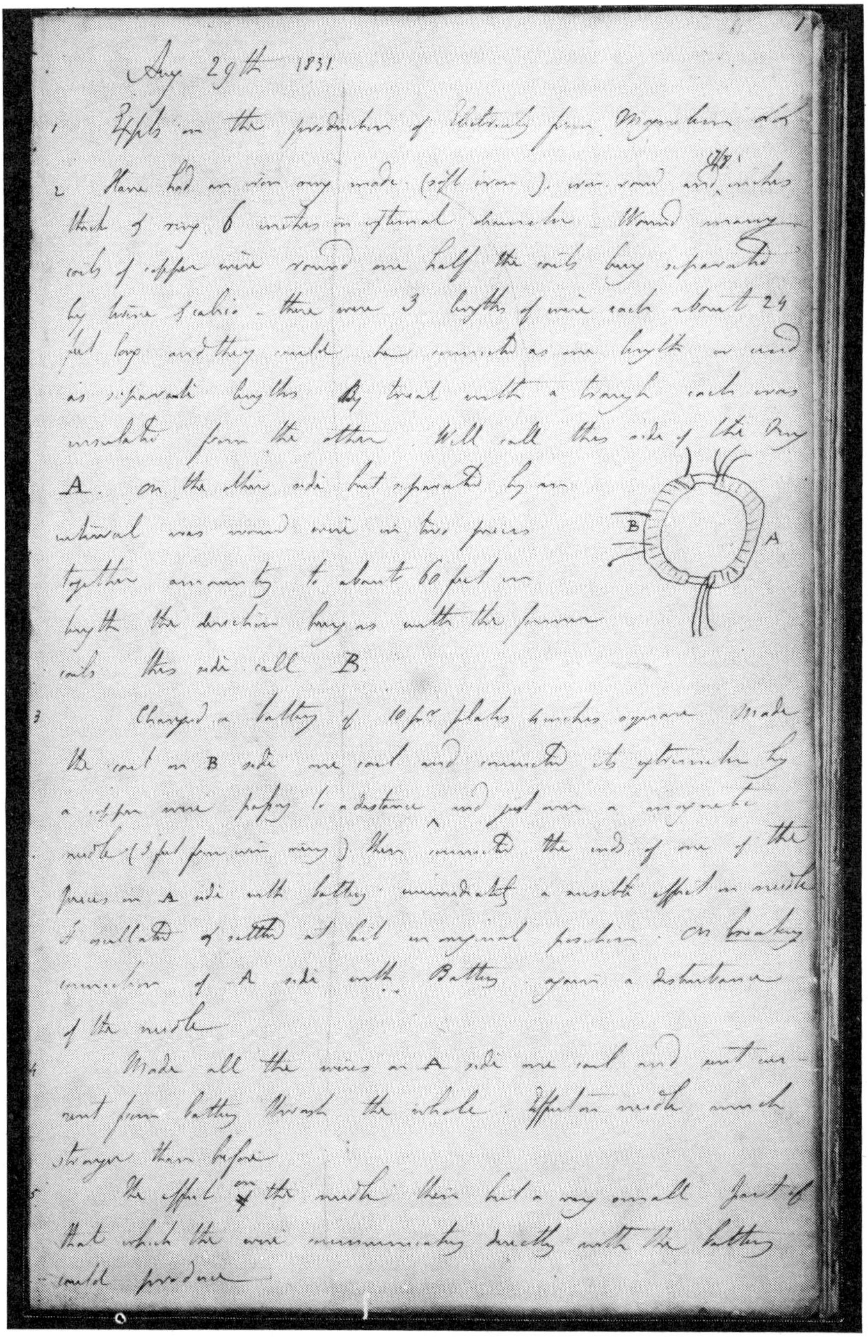

Fig. 2.4 *Entry in Faraday's diary describing his discovery of electromagnetic induction*
Photo: Royal Institution

wound on a cardboard tube.

When the experiment was made without the iron core, it was very like the 1825 experiment. Why therefore had he not found electromagnetic induction in 1825? His experimental arrangements were better in 1831, but his understanding of what he was seeking had changed also. In 1825 he had expected the mere presence of a current in one wire to produce an effect in another. By 1831, possibly influenced by Arago's discovery, he was expecting another factor to be involved. That factor was motion, or a change in something. He expected an effect in the second coil at the moment he completed the circuit of the first coil, and, because he was expecting it, he looked for it and found it.

Fig. 2.5 *Wound iron ring used by Faraday and described in the entry in his diary in Fig. 2.4*
Photo: Royal Institution

Faraday recorded his discovery of electromagnetic induction in his notebook under the heading 'Expts on the production of Electricity from Magnetism &c'. However, he had not produced electricity from magnetism, and this he now sought to do.

He experimented with a variety of coils of wire, magnets, and pieces of iron, and first observed magnetoelectric induction (the production of electricity from magnetism) on 24 September 1831 (Fig 2.6). His method was to arrange two bar magnets and a piece of soft iron in a triangle. The soft iron was surrounded by a coil connected to a galvanometer, and he observed a deflection of the galvanometer needle when the iron was pulled away from the magnets. When it was put back there was another brief deflection in the opposite direction. Faraday noted 'Hence here distinct conversion of magnetism into electricity'. In a further experiment, on

17 October, he produced a current by sliding a cylindrical bar magent into a coil of wire wound on a paper cylinder about 20 mm in diameter and 200 mm long. The coil consisted of eight wires about ten metres long and connected in parallel to a galvanometer.

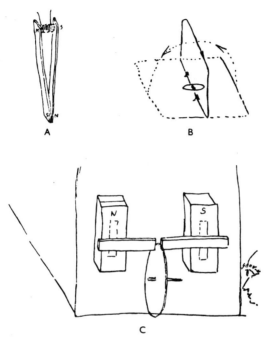

Fig. 2.6 *Sketches from Faraday's diary showing*
    a   the arrangement of coil and magnets he used on 24 September 1831
    b   a loop of wire being rotated in the earth's magnetic field
    c   a copper disc rotating between the poles of the Royal Society's large magnet on
        28 October 1831
    Photo: Royal Institution

Realising that his own magnets were of limited strength, Faraday arranged to use a powerful magnet belonging to the Royal Society. On 28 October 1831 he 'made a copper disc turn round between the poles of the great horse-shoe magnet of the Royal Society'. He connected a galvanometer between metal brushes pressing on different parts of the disc, and found that the galvanometer was deflected as the disc turned. By 4 November he knew that one brush should be at the centre of the disc, or on the axle, and one at the edge.

This series of fundamental discoveries was reported to the Royal Society on 24 November 1831.[2]

The induction effects demonstrated by Faraday appeared to produce 'electricity' identical with the 'electricity' given by frictional machines and by chemical cells, but there was no proof that they were in fact identical. The output of the frictional machines was usually called 'common electricity' and that from chemical cells 'galvanic electricity' or just 'galvanism'. In 1832 Faraday conducted a systematic series of experiments which showed that these electricities, and also 'thermoelectricity', from thermocouples, and 'animal electricity', from electric fish, produced the same chemical, magnetic and other effects.

It was fairly easy to show that the electricity from Faraday's disc generator and the electricity from a thermopile were essentially the same. Both would heat a fine wire, deflect a compass needle, and produce chemical effects such as the decomposition of water. It was much harder to show that static electricity and the discharge from an electric fish were similar to the low voltage effects of the generator and thermopile. Faraday realised that the difference was a matter of 'tension', or voltage. He connected a frictional electric machine to a galvanometer by a damp thread, which he reasoned would have the effect of diminishing the tension, and found that the galvanometer was deflected just as it was by the other electricities. He showed that a frictional machine could produce electrochemical effects by connecting the machine to two silver wires dipped in copper sulphate solution. After turning the machine for some time he found a deposit of copper on one of the silver wires.

2.3   Electrochemistry

One of the experiments Faraday performed in demonstrating the identity of electricity from different sources measured the chemical effect of a given quantity of electricity. He showed that 'The chemical power, like the magnetic force, is in direct proportion to the quantity of electricity that passes'. This was his First Law of Electrolysis. Later he showed that in electrolysis the products were evolved in quantities proportional to their chemical equivalent weights. He expressed this Second Law of Electrolysis as 'Electrochemical equivalents coincide and are the same with ordinary chemical equivalents'. Surprisingly, Faraday did not reach his second law simply by continuing the work that had given the first. It arose instead from his development of the voltameter as a means of measuring quantities of electricity accurately. In his voltameter the current being measured passed through an electrochemical cell containing acidulated water. The hydrogen and oxygen evolved were collected separately, and Faraday observed that the ratio was always precise.

Scientific discussion of electrochemistry was very difficult at the time partly because there was no generally agreed terminology. Faraday consulted the Rev. William Whewell of Cambridge, who had already devised several scientific words. (The two men discussed other scientific words: in 1840 Faraday asked Whewell to think of something better than the new word 'physicist' which, said Faraday, is 'so awkward that I think I shall never be able to use it'). Whewell proposed 'electrolyte' for the liquid, and the terms 'ion', 'cation', 'anion', 'cathode' and 'anode'. Faraday adopted these terms when he published his electrochemical discoveries in January 1834, and they became generally accepted.

The chemistry of electroplating was worked out by several people around 1840, and by 1844 the Birmingham firm of Elkington's was electroplating articles on a commercial scale. They used a massive wooden-framed magnetoelectric machine driven by a steam engine as their source of electricity. A related process which also flourished on an industrial scale was electrotyping. In this small objects, such as coins and medals, were copied by making a wax impression, coating the wax with an electrically conducting layer such as carbon paste, and then electrodepositing metal. The electrolytic refining of metals, especially aluminium and copper, were developed on a commercial scale at the end of the nineteenth century when substantial quantities of cheap electric power became available.

2.4 Faraday's later electrical researches

Throughout the rest of his life Faraday continued to be interested in the fundamental nature of electricity and other physical phenomena. Although interested in the practical application of his discoveries, he rarely became directly involved in their exploitation. One exception, however, was his work on lighthouses. He was Scientific Adviser to Trinity House from 1836 to 1865 and in that capacity visited South Foreland lighthouse for trials of an arc lamp in 1858.

In 1835, when others were developing magnetoelectric generators from Faraday's discoveries of 1831, he turned to electrostatics. He was interested in the nature and the location of the electric charge, and found that whenever a positive charge was produced there was also an equal negative charge. He also showed that when a body was charged the whole charge was on the outer surface. In a dramatic demonstration he sat inside a hollow cube in the lecture theatre. The cube was covered with conducting material and insulated from the floor. A colleague outside charged the cube so highly that sparks flew from it, yet Faraday found that even with his most sensitive equipment he could not detect any electric action within the cage.

Thinking that in a Leyden jar or other capacitor the dielectric (the substance between the conducting plates) must have a bearing on the behavious of the capacitor, he made two identical capacitors with concentric spheres. He showed that the capacities were different for different materials, and that there was a physical property of insulating materials which he called the specific inductive capacity, now usually called the dielectric constant.

He seems to have looked for signs of strain in the dielectric analogous to the lines of force of a magnet, and therefore went on to examine what happens when a dielectric breaks down. This happens most easily when the dielectric is the air, preferably at reduced pressure, between two conductors enclosed in a glass vessel. So he came to study and describe the electric discharge in a partially evacuated tube (Fig. 2.7). He could not explain the phenomena, but noted them carefully, including the dark space which, under certain conditions, occurs near the cathode.

Faraday had an intuitive belief that all the fundamental forces of nature were related. He sought, though never found, a connection between electricity and gravity. In 1845 he did find an interaction between magnetism and light. He found that when polarised light was passed through a block of heavy glass, left over from his optical glass studies of twenty years earlier, the plane of polarisation of the light could be turned by a powerful magnet. His last scientific experiment, in 1862, was to seek a change in the monochromatic light of a sodium flame in a magnetic field. He failed to detect any change. The shift in the spectrum in the presence of a magnetic field is now called the Zeeman effect, after the man who successfully demonstrated it with better equipment in 1896.

Fig. 2.7 *Faraday's 'electric egg' used for studying electrical discharges in gases at reduced pressure*
Photo: Royal Institution

Diamagnetism, the tendency of some materials to set themselves across rather than along the lines of magnetic force, was demonstrated in November 1845, again using heavy glass. His note of this experiment was the first occasion on which Faraday used the term 'magnetic field'.

On 3 April 1846 Faraday gave a Friday Evening Discourse entitled 'Thoughts on ray vibrations'. Tradition has it that Wheatstone was to give a discourse that evening and, being a very shy man, panicked at the last moment, leaving Faraday to fill the gap. The story cannot be proved or disproved, and the Royal Institution lecture programme has no speaker's name for that evening. Whatever the precise circumstances, Faraday spoke at fairly short notice and he expounded thoughts which were only half-formed. He suggested that light might be a disturbance of lines of electric or magnetic force. The idea was received with scepticism and some ridicule at the time, and probably he would not have voiced his thoughts if he had not just shown a relationship between magnetism and light. When in 1864 James Clerk Maxwell expounded his electromagnetic theory of light, he generously acknowledged Faraday with the words:

The electromagnetic theory of light as proposed by him is the same in substance as that which I have begun to develop in this paper, except that in 1846 there were no data to calculate the velocity of propagation.

For several years around 1840 Faraday was in poor health due to overwork, and in 1841 he and his wife spent several months resting in Switzerland. Thereafter his work was a series of short bursts of activity, rather than the prolonged periods of research that had characterised his earlier years. He advised the Government and other public bodies on a number of specific problems. In 1844 there was a public outcry about the safety of coal-mines after an explosion at Haswell Colliery in County Durham, in which 95 miners were killed. The Prime Minister, Sir Robert Peel, decided that a chemist and a geologist should conduct an enquiry into the accident. Faraday and Sir Charles Lyell were chosen, and travelled to Durham to make their enquiry. They discovered that it was possible to light a pipe from a Davy safety lamp and that some men had been smoking in the mine although it was strictly forbidden.

Faraday was one of a committee set up in 1850 to look into the problem of preserving the paintings in the National Gallery. London was a dirty, smoky city and the combined effects of London air and thousands of visitors every day was wreaking havoc on the pictures. The committee recommended that pictures should be glazed, and this was done although they recognised that this was not a complete solution to the problem. Faraday made a study of the chemistry of varnishes suitable for covering pictures. Later he was consulted about the Elgin Marbles in the British Museum. These Greek statues from the Parthenon at Athens had been brought to London by Lord Elgin in 1812, and were deteriorating rapidly in the London air. Faraday reported that nothing could be done to restore the Marbles to their original condition, though he suggested ways of stopping further deterioration.

Prince Albert had a great regard for Faraday and occasionally took his children to hear Faraday lecture. In 1858 he suggested to the Queen that Faraday should be offered a Grace and Favour Residence, a house at Hampton Court. From 1862, having resigned from most of his duties at the Royal Institution, Faraday lived at Hampton Court. He was offered the Presidency of the Royal Institution, but declined because he knew that he was no longer able to carry out the duties. He died on 25 August 1867 and was buried in Highgate cemetery, where the gravestone recorded simply his name and the dates of his birth and death.

The offer of the Presidency was an expression of the regard in which Faraday was held by the Royal Institution. Today he is remembered by electrical engineers as the founder of their technology primarily because of his discovery of induction. The generator and the transformer are central to the modern electrical world and they derive straight from Faraday's work in 1831, yet Faraday himself could not have foreseen the consequences of his 'distinct conversion of magnetism into electricity'.

2.5   References

There are several biographies of Faraday. The most comprehensive is: Pearce Williams, L.: *Michael Faraday*, London 1965

Recent, shorter, biographies are:
King, Ronald: *Michael Faraday of the Royal Institution,* Royal Institution 1973
Bowers, Brian: *Michael Faraday and electricity,* Priory Press, Hove, 1974

1   Faraday, Michael: 'Historical sketch of electro-magnetism', *Annals of Philosophy,* 1822, 3, pp.107-121
2   Faraday, Michael: 'Electro-magnetic researches', *Phil. Trans.,* 1832, pp.125-162

Chapter 3

Telegraphs

3.1 Early telegraph proposals

The first important application of electricity was the electric telegraph, and the first electric telegraph to be used commercially was an arrangement patented by Cooke and Wheatstone in 1837.

A workable electric telegraph was described in an anonymous letter in the *Scots Magazine* of 1753, though there is no evidence that it was ever actually made. It had an electrostatic machine which could be connected to any one of 26 insulated wires. Each wire either ended with a spark gap or was arranged to attract a small piece of paper when charged. The person receiving the message had to see which wire was charged and note the corresponding letter.

Another electrostatic telegraph was actually demonstrated by Sir Francis Ronalds (1788-1873) in the garden of his house at Hammersmith about 1816 (Figs. 3.1 and 3.2). Ronalds was later a founder member of the Society of Telegraph Engineers

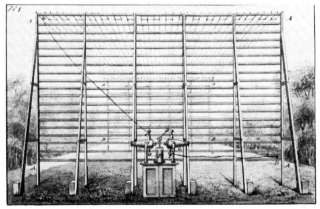

Fig. 3.1   *Ronald's own drawing of his experimental telegraph in the garden of his house at Hammersmith*
(From a drawing in the IEE Archives)

(now the Institution of Electrical Engineers), to which he bequeathed his extensive collection of electrical books. His telegraph used a single wire, partly suspended in the air and partly buried in a glass tube in the ground, which was charged continuously by a frictional electric machine. At both ends a clockwork mechanism turned a dial with letters round it, and the person sending the message earthed the

Fig. 3.2   The electrostatic machine, a piece of wire threaded through a glass tube, and the pith-ball indicator used by Ronalds in his telegraph

line at the instant the desired letter was indicated. The person receiving the message had a pith-ball electroscope so that he could observe the moment at which the line was earthed and note the letter indicated by his dial at that instant. The system was slow and depended on the two dials keeping in synchronism. Ronalds wrote a pamphlet about his system, and worked out a routine for synchronising the dials. Should vandals damage the wire, he advised: 'hang them if you can catch them, damn them if you cannot, mend it immediately in both cases'.

Ronalds offered his telegraph to the Government, but the Secretary of the Admiralty informed him bluntly that no telegraph other than the one then in use would be adopted. The telegraph then in use was a semaphore telegraph linking London and Portsmouth, which enabled the Government to send orders to the fleet in a few minutes — provided it was daylight and not foggy.

S.T. von Soemmering (1775-1830) in Munich demonstrated a telegraph which used current electricity rather than static electricity, but still required a separate wire for each letter. The wires all ended in a vessel of acidulated water, and the signalling current electrolysed the water, producing a stream of hydrogen bubbles at the wire corresponding to the letter being indicated.

Oersted's discovery that a pivoted magnetic needle could be deflected by an electric current led to experimental telegraph systems using an electromagnetic detector. One of these was devised by Professor Müncke, of Heidelberg, who in 1836 gave a demonstration which inspired W.F.Cooke to turn his attention to telegraphy.

3.2 Cooke and Wheatstone

William Fothergill Cooke (1806-1879) was born at Ealing, Middlesex. His father, a friend of Francis Ronalds, was a surgeon and later became Professor of Anatomy at Durham University. Cooke was educated at Durham and Edinburgh, and became an officer in the East India Company's army at Madras, but resigned through ill health in 1833. While convalescent he travelled in Europe and began a new career making anatomical models in wax, but when he saw Müncke's telegraph demonstration in March 1836 he immediately resolved to leave his model-making and devote himself to the practical exploitation of the telegraph.

Cooke's first experimental telegraph used six wires forming three circuits to influence three needles, which permits 26 different indications (each needle may be deflected left or right, or remain at rest). He then made what he called his 'mechanical telegraph' which had the advantage of needing only one circuit. This circuit included an electromagnet which operated a brake on a clockwork driven system. As with Ronalds' telegraph, Cooke's mechanical telegraph required that independent clockwork mechanisms at each end of the telegraph line should run in synchronism. The problem which defeated Cooke at that time, however, was not the problem of keeping two clocks in synchronism but the fact that he could not operate the electromagnetic mechanism through a very long length of wire. He records one experiment in 1836 in which he was unable to signal through a circuit one mile long, although the instruments he was using worked satisfactorily connected through a short wire.

Cooke was alert to the commercial possibilities, and he found a possible customer; he realised that the railways might find the telegraph useful for signalling between stations, and in January 1837 he showed his instruments to the directors of the Liverpool and Manchester Railway. Meanwhile, he continued experiments to find at what distance he could operate his mechanical telegraph, and he sought advice from Faraday. Faraday advised him 'to increase the number of the plates of the battery proportionately to the length of the wires'. This helped, but Faraday did not appreciate that the relationship between the resistance of the electromagnet in the receivers and the resistance of the connecting wire was important. Cooke also consulted Roget, the Secretary of the Royal Society, and Roget referred him to Wheatstone, probably because he knew that Wheatstone had been operating instruments through several miles of wire in the course of some experiments at King's College.

Charles Wheatstone (1802-1875) was Professor of Experimental Philosophy at King's College London from 1834 until his death. He was born in Gloucester, the son of a musical instrument manufacturer and music publisher. He became interested in the working of musical instruments and the physical basis of acoustic phenomena. Wheatstone was a poor public speaker, but Faraday was interested in his work and some of his discoveries were made public by Faraday in Friday Evening Discourses at the Royal Institution. Wheatstone studied the transmission of sound through solid rods and wires and arranged concerts in his father's shop on an 'Enchanted lyre': music played on unseen instruments in an upper room was transmitted along a brass wire to a reconstruction of an ancient Greek lyre, which was all the audience could see.

Wheatstone wondered whether it might be possible to send messages between towns by transmitting sounds through wires; his experiments convinced him, however, that although sound transmission was possible over short distances it was not practical through very long lengths of wire. He then turned his attention to electricity, and set out to discover the speed with which electricity passed along a wire. For this research he had several miles of wire arranged in the basement of King's College, through which he operated instruments. He investigated the conditions for obtaining the maximum effect on the instruments, and this led him to study in detail the measurement of current, electromotive force, and resistance. The two consequences of this work were that he became an authority on electrical measurements (see Chapter 12), and he knew that a telegraph receiver should have the same resistance as the line connecting it in order to achieve maximum sensitivity.

Cooke and Wheatstone ought to have made a splendid partnership. Cooke had the business acumen to see the commercial potential of the telegraph, and he proved a convincing salesman, but his instruments failed because of his own limited scientific understanding. Wheatstone knew how to make a telegraph work, but he did not see that telegraphy could be a profitable business. The first telegraph in commercial use was the five-needle apparatus patented in 1837 and installed on the Great Western Railway in 1838. It was entirely designed by Wheatstone, but Cooke negotiated the agreement with the railway company and supervised the installation.

Sadly Cooke and Wheatstone quarrelled about who should receive the chief credit for their telegraphic achievements. It was a foolish quarrel since their success was due to their joint labours and neither could have succeeded alone. The dispute was submitted to Arbitration in 1840. Cooke wrote a voluminous statement of his 'case' (which is invaluable for telegraph historians) and Wheatstone wrote a 'reply'. In April 1841 the arbitrators, Professor Daniell and Sir Marc Brunel, produced a statement which said little but apparently satisfied the parties. It concluded 'it is to the united labours of two gentlemen so well qualified for mutual assistance, that we must attribute the rapid progress which this important invention has made'.

The arbitration proceedings brought an end, temporarily, to the quarrel between Cooke and Wheatstone. They continued to develop their telegraph ideas and obtained several patents which, according to the partnership agreement, were their joint property. The partners had different ideas about how the telegraph would develop, and each pursued some telegraphic researches independently of the other. Wheatstone thought that a telegraph should be easily operated by unskilled people:

he envisaged its use on private lines where simplicity rather than speed would be important, and his five-needle telegraph met this requirement. Cooke envisaged the growth of bulk telegraph traffic with specialist operators sending coded messages at high speed. As the businessman of the partnership, he also realised that for long distance telegraphy the cost of the wire would be crucial, and they had to have instruments operable through one or two wires, not the five wires of the five needle instrument.

In January 1840 Cooke and Wheatstone obtained a patent including the 'ABC' telegraphs, which were designed to meet Wheatstone's desire for a simple, direct reading instrument. In ABC instruments (Fig. 3.3) letters arranged around a dial on the receiving instrument were indicated successively by a rotating mechanism controlled by the sending instrument (or 'communicator'). To indicate a letter the motion is interrupted briefly at the appropriate point and the person receiving the message notes the letters thus indicated. Cooke's experimental instruments had worked in this way, but he had a local power source driving the dial on the receiver and tried to control it by signals transmitted along the line. Wheatstone designed

Fig. 3.3 *Wheatstone 'ABC' receiver*
A rotating dial carries a ring of letters, one of which can be seen through the window. Each time the dial stops the person receiving the message notes which letter is shown.

instruments with a very sensitive receiver mechanism which was driven directly by the transmitted signal. In 1841 Wheatstone made a printing ABC telegraph (Fig. 3.4). The receiver had what would now be called a 'daisy wheel printer' in place of the dial (Fig. 3.5). This was a wheel with twenty-four springy arms each carrying a

Fig. 3.4 *Wheatstone's printing telegraph of about 1840*
Letters are printed on a sheet of paper wrapped around the horizontal drum. After each letter the drum is turned slightly on the screwed rod that supports it.

printing type. The ABC mechanism turned the wheel until the appropriate letter was in position and then a separate circuit was used to trigger a mechanism which struck the type onto the paper through a piece of carbon paper. A drum which carried the paper advanced automatically after each letter. It was an ingenious machine, and really worked, though was never brought into commercial use.

None of the ABC telegraphs developed around 1840 were a commercial success then, though twenty years later Wheatstone designed improved versions which sold in large numbers. During the 1840s a market developed for simple telegraphs for use on the railways.

The five-needle telegraph which the Great Western Railway had let Cooke instal in 1838 operated between Paddington and West Drayton. In 1842 Cooke obtained permission to extend the telegraph to Slough, but using simpler instruments with only two needles. In 1843 it was extended to Windsor, from where the news of the birth of Queen Victoria's second son was transmitted to London on 6 August 1844. *The Times* acknowledged that it had received the news by 'the extraordinary power of the Electro-Magnetic Telegraph'. Also in 1844 the Admiralty, which had so flatly rejected Ronalds' plans thirty years earlier, entered into a contract with Cooke for a telegraph between the Admiralty in London and Portsmouth. Further publicity for the telegraph came in January 1845 when the murderer John Tawell was seen to board a train at Slough. A description was telegraphed to Paddington where the police met the train.

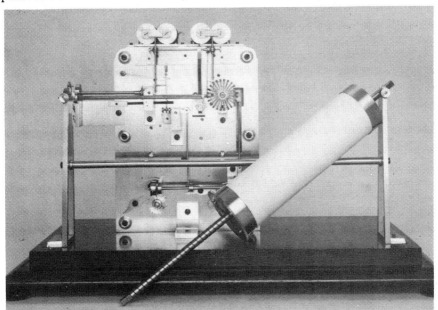

Fig. 3.5 *The printing telegraph with the drum removed to show the type wheel*
A hammer behind the type wheel strikes the appropriate letter on to the paper.

The technical problems associated with the telegraph were gradually solved, but when it began to be exploited commercially a new problem arose. The partners had already financed the construction of the basic telegraph instruments, but the purchase and installation of the wires was very expensive. The telegraph needed far more capital than Cooke and Wheatstone could raise as a partnership, even if Wheatstone had wished to remain an active business partner.

In April 1843 Wheatstone and Cooke negotiated a new agreement between them under which Wheatstone assigned his share in their patents to Cooke in exchange for a royalty payment based on the length of telegraph line completed each year. The agreement included a statement that it was made in order to simplify the business arrangements: up till that date all contracts had to be approved by both partners. All outstanding claims of one partner against the other were cancelled and the royalty was fixed at £20 per mile of telegraph completed during the year for the first ten miles, reducing in stages to £15 per mile after fifty miles.

Cooke's records give some idea of the sums involved. Between 1836 and 1844 he paid out £31 000 and in 1845 alone £71 000. During that period he received £97 000. It was clear that a Company would have to be formed to raise the necessary capital for further telegraph lines. With this object in view Cooke sought financial backing to buy Wheatstone out completely. In August 1845, when the formation of a company was being considered but before Cooke had a financial backer, Wheatstone stated that he would commute his royalty on all lines in England, Wales, Scotland, Ireland and Belgium for a total sum of £30 000. This was agreed after some discussion, together with another agreement which permitted Wheatstone to use the patented apparatus freely on lines not exceeding half a mile in length.

Cooke obtained the support of John Lewis Ricardo (1812-1862), a financier and Member of Parliament for Stoke on Trent, whom he first met on 1 October 1845. It appears that Wheatstone and Ricardo did not meet until 15 January 1846. Ricardo was very impressed with the telegraph and soon came to an agreement with Cooke. He was an able administrator and took a leading part in the promotion of the Electric Telegraph Company, to which the patents were assigned in 1846. Ricardo was chairman for ten years and during that period he introduced the use of franked message papers and the employment of female clerks. He was also concerned with a number of railway companies.

The Electric Telegraph Company was incorporated by Act of Parliament in 1846, and Cooke transferred his interests to the new company for £150 000. There was a suggestion that Wheatstone should be a scientific adviser to the company, but that was not implemented.

Cooke and Wheatstone had no further formal business connections. Cooke was a director of the Electric Telegraph Company for most of his life. He resigned from the Board in November 1849 after a disagreement, but was re-elected later. He made no more inventions in connection with the telegraph, and later he became interested in quarrying and stone cutting machinery, obtaining a series of patents. He moved to Wales and acquired a quarry in Merionethshire, but his new interest proved a financial disaster and he lost all the fortune the telegraph had brought him.

Wheatstone engaged the services of several scientific instrument makers in the course of his telegraph works. One such at this time was W.T.Henley (1814-1882), a self-taught scientist who set up in business as a scientific instrument maker and eventually founded W.T.Henley's Telegraph Works Co. Ltd.

In 1843 and 1844 Wheatstone conducted some underwater telegraphy experiments. The first were in the Thames, between King's College on the north bank and a building on the south. The 1844 experiments were conducted in Swansea Bay, when Wheatstone was staying with his friend L.W.Dillwyn, a Member of Parliament and a Fellow of the Royal Society. It is not clear why Wheatstone travelled so far to conduct the experiments, but one of Dillwyn's neighbours owned a copper works and the cable for the trials could have been manufactured there. Wheatstone tried underwater cables insulated with rope soaked in tar and also worsted and marine glue (Fig. 3.6). Neither was a success, however, and practical submarine telegraphy had to await the introduction of gutta percha insulation extruded without any joint on to the conductor.

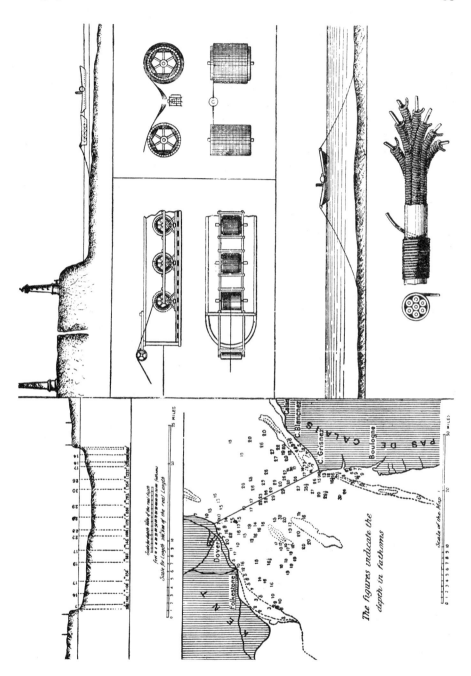

Fig. 3.6 *Wheatstone's drawings showing how a submarine cable might be constructed and laid between England and France*

3.3   Morse

At the same time as Cooke and Wheatstone were developing their telegraphs in Britain, Samuel Findlay Breeze Morse (1791-1872) was working on a telegraph in the USA. Morse was neither a scientist nor an engineer by training, but a painter and sculptor who had gained renown in both Europe and America. In 1812 the Society of Arts in London awarded him a gold medal for his work.

In 1832, while homeward bound from Europe, Morse met another passenger who had heard lectures on electromagnetism in Paris. In the course of many conversations Morse conceived the idea of an electric telegraph. From the start he envisaged signalling by an intermittent current using a code. He gradually abandoned his art in order to give time to the telegraph, but was fortunate to be appointed the first Professor of the Arts of Design at the new University of New York, which gave him a regular income.

Morse sought the help of two other men. Professor Gale, a chemist, helped him with his battery and apparatus. Alfred Vail, a man with a mechanical turn of mind, helped with the manufacture of equipment and also devised the code of long and short signals known ever since as 'Morse code'.

Morse obtained a United States patent for his telegraph on 3 October 1837. In June 1838 he sailed to England to seek an English Patent, but the application was opposed by several people, including Cooke and Wheatstone, on the ground that Morse's work had all been published previously in England, in the *Mechanics Magazine* of 10 February 1838. Morse seems not to have known that prior publication of an invention made it impossible to get a patent in England, and he regarded it as most unjust when the Attorney General refused his application. Although Morse failed in the main object of his visit to London, he found it useful to meet other telegraph workers. He also found some consolation in the fact that while in London he managed to obtain a seat in Westminster Abbey to see Queen Victoria's coronation on 28 June 1838.

Morse persuaded the United States Congress to let him erect a telegraph line between Baltimore and Washington. The line first worked in 1843 and opened as a public service on 1 April 1845. Thereafter Morse quickly gained financial backing, and telegraph circuits spread throughout the USA. New York and San Francisco were linked in 1861.

Morse's original receiver worked by making indentations in a paper tape, but he then made a machine which produced an ink trace. Some of his operators discovered that, with experience, they could 'read' the Morse signals by listening to the clicks of the instrument, without looking at the tape. Morse himself did not approve of the practice.

3.4   The exploitation of the telegraph

By 1846 the electric telegraph had proved its worth and it was about to embark on a period of rapid growth. Governments, business, the press, and the public in general all appreciated its potential value. In most European countries the telegraph was developed as a state monopoly, but in Britain, and also in the USA, it was left to private enterprise.

The Electric Telegraph Company, already mentioned, had an authorised capital of £600 000, though only £112 00 was paid up initially. By November 1848 telegraphs covered 1800 miles of railway, which was about half the railways then open. The telegraph from London to Bristol was completed in 1852. Most of the telegraph lines were laid along the railways, and the railways provided the telegraph companies with the majority of their business. When the telegraphs were nationalised under an Act of 1868 a major practical problem was the large number of agreements and working arrangements between the telegraph and railway companies. In taking over the telegraphs the Post Office did not wish to incur a liability to provide telegraph services for the railways, but many telegraph lines had been established on the understanding that the railway companies could always get their messages through quickly.

By April 1850 the railway system had grown to over 7000 miles, though the Electric Telegraph Company had installed only 2200 miles of telegraph. Not surprisingly, other companies set up in competition, the first being the English & Irish Magnetic Telegraph Company, formed in August 1851. In November 1851 the first international submarine cable, between Dover and Calais, was brought into service by yet another company, the British Electric Telegraph Company. Two other companies are worth special mention here. The London District Telegraph Company was formed in 1859 to exploit the potential traffic close to the capital. The Universal Private Telegraph Company was formed in June 1861 to exploit Wheatstone's later ABC telegraphs. This company, which had a capital of £190 000, offered private lines to customers using ABC telegraphs. Unlike the other companies it did not itself convey messages. It provided a private line between any two points, such as home and business, for £4 per mile per year.

In 1868 there were 80 000 miles of telegraph wire in the United Kingdom, and more than 5 600 000 messages were sent. In addition almost 800 000 messages were sent overseas, to Europe and to America, using the cable opened in 1866. The usual charge for sending a message up to 100 miles in 1868 was one shilling (5p) for 20 words and up to twice that for longer distances. The basic charge introduced by the Post Office in February 1870 was one shilling for 20 words for any distance within the UK, and the price did not change greatly until 1915.

The plot of Thomas Hardy's novel *A Laodician,* first published in 1881, depends on its heroine's familiarity with the telegraph. A telegraph historian reading Hardy's book may object that Hardy was confusing the public telegraph service between telegraph offices and the ABC telegraphs on private wires. Nevertheless, it is noteworthy that by 1881 the electric telegraph had become part of the English country scene, familiar to everyone even though it was too expensive for most people to use except on very special occasions.

As previously mentioned, most of the European telegraphs were state monopolies from the beginning, though the construction work was usually undertaken by private firms. In 1846 the Prussian authorities were investigating the possibilities of the telegraph. A man closely involved in this work was a young army officer, Werner Siemens (1816-1892).

Siemens was born at Lenthe, near Hanover, the eldest son of a tenant farmer. He wanted a scientific education, and the only way that could be achieved without money was by joining the army. He spent three years at the Artillery and Engineering College in Berlin, leaving as a lieutenant in 1838 and joining the Royal Artillery Workshops. Shortly after both his parents died and Werner assumed responsibility for his younger brothers. He envisaged, even at that early date, the development of a family business with each brother having a distinct sphere of responsibility. As part of this plan the second brother, Carl Wilhelm (1823-1883), went to England in 1844, and settled permanently. He became a British subject, anglicising his name to Charles William Siemens. He was later knighted and known as Sir William Siemens. Another brother, Carl (1829-1906), settled in Russia from 1853.

The firm of Siemens & Halske was formed in 1847. Johann Georg Halske (1814-1890) was a university technician whom Werner Siemens had met and subsequently employed. Initially Halske managed the business since Siemens was still a serving army officer, but he soon resigned his commission to give his full attention to the business. About this time he told his brother in England

Telegraphy will become a separate branch of scientific technology, and I somehow feel a call to set it going, as it is still, I am convinced, only in its infancy.

For some years Werner Siemens was occupied supplying telegraphs for the Prussian Government. The business flourished, both in Germany and in the British and Russian branches. In time they branched out into virtually every aspect of electrical engineering, with a network of trading contacts and manufacturing premises covering most of Europe. By 1900 Siemens & Halske was one of the largest electrical engineering concerns in the world.

3.5   Technical progress in telegraphy

The telegraph instruments with which the large scale telegraph business was founded were quite simple. The sender was a manually operated 'Morse' key; the receiver was a single needle to be watched, a sounder, or a Morse inker or similar instrument which recorded the signals on a paper tape. All worked at the speed human operators could manage, but no faster.

There was a powerful incentive to send messages faster. It was not that the speed of transmission of one message mattered, but that more messages could be sent over one line. The actual telegraph wires were the most expensive part of the telegraph system, and it was therefore important to use the wires as efficiently as possible. The system adopted was that several clerks in a telegraph transmitting system should punch the messages on to a tape, using Morse or similar code. Several clerks would be working preparing tapes simultaneously which were then transmitted automatically through a single set of equipment which operated far quicker than a human operator could either transmit or receive the signals.

A number of automatic systems were tried by the telegraph companies, all using punched paper tapes. The chemical printer developed by Alexander Bain was the receiver used from the early 1850s. This used paper soaked in a colourless chemical which was decomposed electrolytically by the signal current to yield a coloured product. Potassium iodide was generally used, which yields iodine. Subsequently receivers were adopted in which a wheel fed with ink was pressed on the paper when a signal was received. The best results were obtained when a magnetically polarised armature operated the inker and the Morse signals were currents of one polarity while a current of the opposite polarity was transmitted in the spaces between the 'dots' and 'dashes'. This proved more satisfactory than a signal current which simply went on and off.

In most automatic transmitters springy contacts pressed on the paper tape and a circuit was completed when the contacts met through holes in the tape. The maximum speed of operation of the system was determined by the transmitter; the printers at the receiving end could always keep up. The limiting factor was that either the paper tape tore or that the contacts failed to close through the holes if the tape was driven too fast. The most important of Wheatstone's telegraphic inventions was his automatic transmitter, developed first in 1858 (Fig. 3.7). In this the tape was driven by a sprocket engaged in a continuous line of holes while the message was indicated by the presence or absence of holes in two positions adjacent

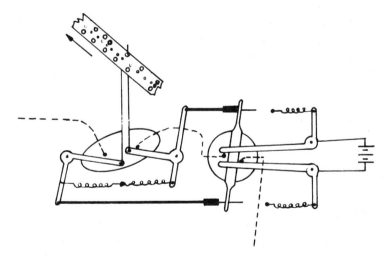

Fig. 3.7 *Principle of the Wheatstone automatic telegraph*
At each position the vertical 'feeler' rods on the left are raised and either pass through a hole in the paper or are restrained by the paper. The linkage turns the disc on the right to connect the telegraph wires (shown dashed) to the battery in the appropriate sense. There is always current in the line: the mechanism sends signals by reversing the direction of the current.

to the drive holes. There were no contacts feeling through the tape, but at each position two light 'feeler' rods were pressed by a spring against the positions in the tape where the holes would be. The mechanism detected whether each feeler rod was able to pass through the paper at that point, or whether it was restrained because there was no hole, and operated contacts accordingly. The paper itself did not have to move the contacts or any other mechanism, but simply restrained the feelers or let them pass as appropriate. Consequently, there was no stress on the paper except the steady force pulling it along, and the system could operate much faster than earlier arrangements. The code Wheatstone used was basically Morse code but he did not use dashes. The dot of Morse code was represented by two holes in line across the tape; the dash was represented by two holes staggered, one a space ahead of the other and on opposite sides of the tape. According to W.H.Preece, later Sir William Preece, Engineer-in-Chief of the Post Office, Wheatstone's first automatic telegraph could operate at 70-80 words per minute. With the various refinements Wheatstone made later the speed was increased to 180-190 words per minute. This meant that the message-carrying ability of the telegraph system was more than doubled without any extra wires being laid, and in Preece's view the prolific growth of telegraph business in the 1870s was largely due to Wheatstone's apparatus.

At the same time as Wheatstone was developing his automatic telegraph transmitter for use on the busiest commercial lines he was also perfecting an instrument for a very different market. This was his 'ABC' telegraph, for use by unskilled people on private lines. It was really a much improved version of the earlier ABC instruments he had made around 1840, and the 1860s version proved a commercial success. The technical refinements alone, however, cannot explain why the instruments sold well in the 1860s but not twenty years earlier. The explanation must lie in the new public understanding of the electric telegraph: people knew that messages could be sent over the wires, and wealthy people wanted to buy the wherewithal to send messages between home and business, or between offices. Wheatstone's Universal Private Telegraph Company mentioned above was formed to supply this want, and Wheatstone himself was free to attend to the practical problems of perfecting the ABC system. Around 1840 his relationship with Cooke, and other business matters too, had made it difficult for him to devote much time to this kind of telegraph.

3.6   The fruits of the telegraph

The development and practical introduction of the electric telegraph brought enormous changes to the world, but this is not the place to discuss the social and commercial impact of the fact that information could be transmitted round the world in seconds, rather than being restricted to the speed of boat or horse. Less obvious, but equally important, was the impact of the electric telegraph on electrical science and technology. The terms 'electrical engineer' and 'electrical engineering' were not in general use until the end of the nineteenth century (the pioneers called themselves 'electricians'), but the science, technology, industry and profession of electrical engineering all grew out of the telegraph.

Much of the early scientific understanding of electric circuits arose from the study of telegraph operation. Wheatstone's paper on electrical measurements presented to the Royal Society in 1843 (see below, Chapter 12) was the first attempt at a comprehensive survey of the subject, and he said in his introduction

> The practical object to which my attention has been principally directed . . . was to ascertain the most advantageous conditions for the production of electric effects through circuits of great extent, in order to determine the practicability of communicating signals by means of electric currents to more considerable distances than had hitherto been attempted.

The telegraph pioneers worked out such practical matters as how to insulate wires and run them from point to point; when wires were needed later for electric lighting the necessary practical knowledge and experience was already available. Similarly the manufacturing industry which could manufacture cables, coils, electromagnets, instruments and mechanisms was already in existence and could be adapted to the requirements of electric lighting.

The professional organisation, the Society of Telegraph Engineers, was formed in 1871 with C.W.Siemens as its first President. At the Society's opening meeting on 28 February 1872, C.F.Varley, an experienced telegraph engineer, prophesied that 'This Society will . . . develop more into an electrical society than into a society of telegraphy proper'. By 1879 it was taking an interest in electric lighting, and in 1880 the title was broadened to the Society of Telegraph Engineers and of Electricians. The present title, the Institution of Electrical Engineers, was adopted in 1888.

Many pioneers of electricity supply first entered the electrical engineering profession as telegraph engineers. Two such were Sir William Thomson and Thomas Alva Edison.

William Thomson (1824-1907) was born in Belfast but spent most of his childhood in Glasgow where his father was Professor of Mathematics at the University. He distinguished himself at school and at Cambridge University, and then at the age of 22 was appointed Professor of Natural Philosophy at Glasgow. He was a scientific adviser to the Atlantic Telegraph Company and received a knighthood when the Atlantic was successfully crossed by telegraph in 1866. He received many honours, was President of the Royal Society, and three times President of the Institution of Electrical Engineers. In 1892 he was raised to the peerage as Baron Kelvin of Largs, the first man to be ennobled for scientific achievement.

Thomson's most important invention was his siphon recorder, used as a receiving instrument on the Atlantic telegraph. Initially the Atlantic Telegraph Company used mirror galvanometers as their receivers, since they were much more sensitive than the needle instruments used for inland telegraphy. The siphon recorder was even more sensitive than the mirror galvanometer and had the advantage of making a permanent record, while the mirror galvanometer had to be watched continuously by an operator.

The siphon recorder consisted of a light rectangular coil of wire suspended between the poles of a powerful electromagnet so that a current in the coil would cause it to rotate, in the same way as the coil in an ordinary galvanometer. A very fine glass tube, bent to form a siphon, was suspended so that the upper end rested in a vessel of ink or other coloured liquid and the lower end rested on a strip of paper, and could move across it. The coil was connected to the siphon by silk threads in such a way that any twisting of the coil moved the siphon across the paper. The paper was pulled steadily through the instrument, and a wavy line was drawn on it corresponding to the signals in the coil. In some of the instruments Thomson electrified the ink so that it was attracted to the paper. The siphon did not have to touch the paper, so that friction was reduced and the sensitivity increased.

Thomas Alva Edison (1847-1931), the American inventor, began work as a newsboy on a railway. He quickly expanded the job and published his own paper. He learnt to be a telegraph operator and devoted much of his income to experiments. At the age of about 20, he opened his own experimental workshop in Boston. One of his inventions then was a vote recording machine, but he could find no purchaser and ran into debt. He resolved never again to work on an invention for which there was no market. In 1869 he moved to New York, and the following year he set up in business as a telegraph instrument manufacturer in Newark, New Jersey. In his spare time he made further telegraph inventions and also invented his mimeograph and his 'electric pen' for cutting wax stencils, but Edison's ambition was to be a full-time inventor.

In 1876 Edison gave up manufacturing and established a laboratory at Menlo Park, New Jersey. There, with a group of friends and assistants, he spent the happiest years of his life, devoting his whole attention to research and to the development of his inventions. He worked long hours, often neglecting his family and having meals sent over to the laboratory from his house. The phonograph and Edison's filament lamp (see Chapter 8) were developed at Menlo Park, and so was his carbon microphone for the telephone. Alexander Graham Bell (1847-1922), a Scotsman working in Canada and the USA, was developing his telephone through the 1870s just ahead of the development of practical filament lamps. Bell's first telephone used an electromagnetic transmitter and he and others had the idea that a better transmitter might be made using a variable resistance that could be influenced directly by sound waves. In 1877 Edison made his carbon microphone which utilises the fact that the resistance between two pieces of carbon in contact depends on the pressure between them.

The telephone and later developments in telegraphy are outside the scope of this book, but the early progress of telegraphy is the early history of electrical engineering. The events outlined in this chapter are the background from which electric light and power have developed.

3.7   References

A good, comprehensive account of early electric telegraphs is given in Fahie, J.J.: *History of the electric telegraph to the year 1837,* London 1884

For Wheatstone see Bowers, Brian: *Sir Charles Wheatstone,* Science Museum and HMSO, 1975. Much of the correspondence between Cooke and Wheatstone and the arbitration papers (some of which have been published) are in the Archives of the Institution of Electrical Engineers, London

For Morse see Morse, E.L.: *Samuel F.B.Morse: his letters and journals,* Boston and New York, 1914

For the commercial history of the telegraph see especially Kieve, Jeffrey: *The electric telegraph − a social and economic history,* David and Charles, Newton Abbott, 1973

Chapter 4

Early electric motors

4.1 Motion from electricity

The origin of electric motors may be traced back to Oersted's discovery of the deflection of a compass by the electric current and Faraday's production of continuous motion by electromagnetism. During the nineteenth century many 'electromagnetic engines' were designed and made but, despite the hopes of their promoters and the support of the technical press, none achieved lasting success as a source of power. However, the research and development work that went into these machines increased the understanding of electrical technology and prepared the way for practical motors.

After Faraday's rotators the next machine in sequence was Barlow's wheel, also of 1821. Peter Barlow (1776-1862) was Professor of Mathematics at the Royal Military Academy, Woolwich. His star-shaped wheel was mounted on a horizontal axis between the poles of a horse-shoe magnet. The wheel dipped into a pool of mercury. When a current was passed between the axis of the wheel and the mercury, then the wheel rotated.[1]

Faraday's device and Barlow's wheel demonstrated that it was possible to produce continuous motion by electrical means, but they had the fundamental limitation that only a single current-carrying conductor was passing through the magnetic field. In 1832 the French instrument maker Hippolyte Pixii (1808-1835) made a generator (though he did not call it that) in which there was relative rotation between a permanent magnet and a *coil* of wire. Pixii later fitted a crude cam-operated reversing switch which performed the function of a commutator and made the output unidirectional.[2] (Pixii's machine is discussed further in Chapter 6, and shown in Fig. 6.1).

Before tracing the development of electromagnetic engines, one idea from the 1820s is worth mentioning in a digression. In 1824 the *Mechanics Magazine* described a device to show the 'Mechanical effects of electricity'. A light wheel with vanes cut out of paper was mounted on an axle beneath two point electrodes in such a way that the vanes could move in the space between the electrodes from one electrode to the other. When an electric discharge from a frictional electric machine or a Leyden jar was passed between the electrodes, then the wheel turned, with the vanes moving from the positive to the negative.[3]

Electromagnets were studied by several people, one of the first being William Sturgeon (1783-1850) who lectured at the East India Company's Academy, near London. Gerard Moll, Professor of Natural Philosophy in the University of Utrecht, was in England in 1828 and saw some of Sturgeon's experiments. Moll then obtained an electromagnet and made experiments relating the weight supported by the electromagnet to the active area of zinc in the battery supplying the current. He also tried to observe the speed with which the magnetism could be created and destroyed when the circuit was made and broken, and the rapidity with which the polarity could be reversed.[4]

In 1830 Professor Salvatore dal Negro of Padua University in Italy obtained rotary motion from an electric current by means of an electromagnet. A permanent magnet hanging from a pivot was attracted to the electromagnet and the electromagnet was energised through a contact which was broken when the permanent magnet moved from its rest position. In this way the permanent magnet was made to perform an oscillatory motion which was converted into rotary motion by a pawl and ratchet.[5] This has been called the first electric motor, but the title seems rather undeserved.[6]

The first person to appreciate that electromagnetism might be used to provide mechanical power was probably the American Joseph Henry (1797-1878), who wrote in 1831[7]

I have lately succeeded in producing motion in a little machine by a power, which, I believe, has never before been applied in mechanics — by magnetic attraction and repulsion. Not much importance, however, is attached to the invention, since the article, in its present state, can only be considered a philosophical toy; although, in the progress of discovery and invention, it is not impossible that the same principle, or some modification of it on a more extended scale, may hereafter be applied to some useful purpose.

A remarkable prophecy!

Henry's engine consisted of a straight electromagnet mounted horizontally and supported on a knife edge at its centre. A bar permanent magnet was placed vertically below each end of the electromagnet with the north poles facing in the same direction. The electromagnet coil was connected between two pairs of stiff wires, one pair at each end, and these wires dipped into mercury cups when their end was down. A battery was connected to each pair of mercury cups. When one end of the bar was depressed, the electromagnet was energised in such a way that the magnetic reactions between the poles of the electromagnet and those of the two permanent magnets caused an attraction at one end of the bar and a repulsion at the other, so that the bar rocked in the other direction. This movement shifted the connections of the coil from one battery to the other and the direction of current in the coil was reversed; the forces acting on the bar were then reversed and the bar returned to its original position. Thus a reciprocating motion was established.

Henry's dipping contacts performed the function of the commutator in a conventional motor. According to Henry's own account the horizontal electromagnet was seven inches (18 cm) long and wound with three parallel strands of 'copper bell-wire', each 25 feet (7·6 m) long. (By 'bell-wire' Henry meant wire used for pulling mechanical bells). The machine ran uniformly at about seventy-five oscillations per minute.

4.2   Electromagnetic engines

Many designs of electromagnet engines appeared in the 1830s and 1840s. One made by Sturgeon in 1832 (Fig. 4.1) embodied his invention of the commutator as we know it today, and it was the first electromagnetic engine to do useful work — it turned a roasting spit![8]

Fig. 4.1   *Reconstruction of Sturgeon's motor of 1832*
The vertical shaft carries two compound permanent magnets, one at the top and one at the bottom, with their like poles pointing in opposite directions. Two concentric wooden cups filled with mercury are mounted on the shaft, and turn with it. Four terminals are fitted to the exterior of the outer cup, two of which are connected to the mercury in the inner, and two to that in the outer cup. From each terminal a piece of wire hangs down and makes contact with one of four fixed quadrant-shaped plates, which are separated from each other by narrow radial gaps. Four vertical soft-iron cylinders are wound with copper wire and placed just clear of the magnets. The lower ends of each set of coils are connected to the adjacent quadrant. Wires from the upper ends dip into the inner mercury cup, and the supply wires dip one into each cup.

Another early but little known maker of an electromagnetic engine was Sibrandus Stratingh (1785-1841), a doctor of medicine and professor of chemistry at Gröningen, Holland. He wanted to make an electric car. In 1835 he made a table model which ran until the battery was exhausted, but he never made a full size road vehicle. He did, however, make an electric boat, in which he took his family in 1840. He was assisted in all this by an instrument maker, Christopher Becker. From contemporary drawings (Fig. 4.2) of the model it appears that Stratingh's and Becker's motor had a straight horizontal rotor, pivoted on a vertical axis at its centre, and a semicircular stator. Both were wound, and there was a simple commutator. The rotor axis was geared to one wheel of a three-wheeled trolley carrying both the motor and a single voltaic cell.[9]

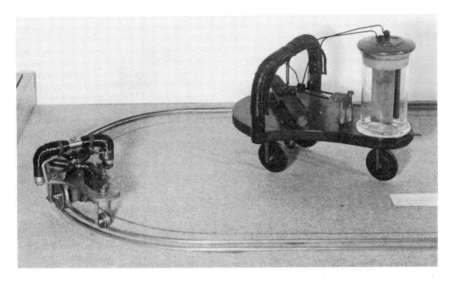

Fig. 4.2 *Two views of Stratingh and Becker's table-top electric vehicle, 1835.*

The earliest patent for electromagnetic engines was obtained by Thomas Davenport. He had suggested using electromagnetism to produce motion in 1833, and he patented a machine in the United States of America in 1837.[10] In the same year an English patent agent, Miles Berry, acting on behalf of Davenport, obtained a similar patent in England.[11] Davenport's patent specification gives a remarkably well developed description. The American Patent Office required a model, and that model is now in the Smithsonian Institution in Washington. Davenport's rotor consists of four coils on a cruciform frame fixed to a vertical shaft (Fig. 4.3). Opposite coils are connected in series and the ends are brought out below the coils as rudimentary brushes, which turn with the rotor. The fixed part of the commutator is two semicircular copper pieces, connected to the battery. The stator is two semicircular permanent magnets, with their like poles adjacent. The specification envisages the possibility of a four (or more) pole stator, in which case the fixed part of the commutator requires correspondingly more segments. The stator magnets could be electromagnets rather than permanent ones.

Courtesy Smithsonian Institution

Fig. 4.3   *Davenport's motor of 1837*
(Photo: Smithsonian Institution)

Davenport was apparently a blacksmith in Rutland, Vermont, though his patent agent just described him as a 'gentleman'. In 1837 he employed a motor weighing fifty pounds (23 kg) and running at 450 revolutions per minute to drill holes a quarter of an inch (6 mm) in diameter in iron and steel. Davenport had great hopes for future electrical developments:

> I hope not to be considered an enthusiast, when I venture to predict that soon engines capable of propelling the largest machinery will be produced by the simple action of two galvanic magnets, and working with much less expense than steam![12]

The London instrument maker Francis Watkins designed a motor in 1835 which he said was inspired by Saxton's generator. Thus it seems that Watkins was recognising that motors and generators are basically the same machines. In reporting on the subject to the Royal Society of London he used his motor to drive models of hammers and pumps.[13]

A typical machine of the 1840s is W.H. Taylor's motor which was enthusiastically written up by the *Mechanics Magazine*[14] (Fig. 4.4). Taylor was an American who had moved to London and obtained an English patent for his motor. This was a simple arrangement of four electromagnets on a frame surrounding a wooden wheel with seven soft iron armatures on the periphery of the wheel. A simple commutator on the axis switched the four electromagnets sequentially. Taylor claimed that previous plans for electromagnetic engines had depended on changing the polarity of electromagnets and that his invention was the idea of switching them so that they were 'alternately and (almost) instantaneously magnetized and demagnetized without any change of polarity whatever taking place'. Taylor's wording is not very clear, but he seems to have realised that earlier engines may have failed because it took a significant time to reverse the polarity of an iron-cored electromagnet, whereas switching it was an almost instantaneous process.

Mechanics' Magazine,

MUSEUM, REGISTER, JOURNAL, AND GAZETTE

No. 874.]	SATURDAY, MAY 9, 1840.	[Price 3*d*.
	Printed and Published for the Proprietor, by W. A. Robertson, No. 166, Fleet-street.	

TAYLOR'S ELECTRO-MAGNETIC ENGINE.

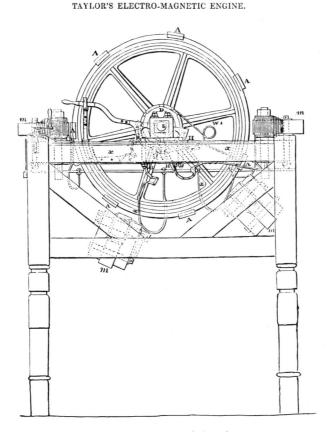

Fig. 4.4   *Report of Taylor's motor in the Mechanics' Magazine*

Taylor's claim to novelty was quickly disputed. Professor P. Forbes of King's College Aberdeen wrote a letter to Faraday which, when published,[15] first brought the work of the Scotsman Robert Davidson into the public eye. According to Forbes, Davidson had 'employed the electro-magnetic power in producing motion by simply suspending the magnetism without a change of the poles' in 1837. Davidson and the Russian Jacobi are the two pioneers of electric motors whose names are best known today.

Forbes did not describe Davidson's motor in detail, but said it had only two electromagnets. It could drive a lathe and turn small articles when supplied from a battery having one square foot (0·09 m^2) of zinc surface. Another machine powered by the same battery would propel a small carriage with two people along the wooden floor of the room.

Professor Moritz Hermann Jacobi of St Petersburg obtained a grant from the Czar to enable him to conduct research into electric power — the first government grant for electrical engineering research. He described his work with an electrically driven boat in a letter to Faraday in 1839. Jacobi's boat was a ten-oared shallop fitted with paddle wheels driven by an electromagnetic engine. He 'travelled for whole days', usually with ten or twelve people aboard, on the River Neva. Jacobi's motor consisted of two wooden frames, each carrying a series of horse-shoe electro-magnets with pole pieces facing one another but spaced apart. The rotor moved in the space between the pole pieces. It had a six-legged spider with a straight permanent magnet on each leg. A simple commutator periodically reversed the connections between the battery and the coils so that the rotor would always move in the same direction. When supplied through a battery of 128 Grove cells the vessel travelled at just over four kilometres per hour. Jacobi reported that the motor could provide 'a force of one horse' from a battery with twenty square feet (1·9 m^2) of platina. Another motor by Jacobi had electromagnets on both stator and rotor[16] (Fig. 4.5).

In the winter of 1841-2 the Royal Scottish Society of Arts gave Davidson financial help to continue his experiments. In September 1842 he tried an electrically driven carriage on the Edinburgh and Glasgow Railway; the four-wheeled carriage was 16 feet long (4·9 m) and weighed about five tonnes (Fig. 4.6). The motors consisted of wooden cylinders on each axle with iron strips fixed in grooves in the cylindrical surface. Horse-shoe magnets one either side of the cylinder were energised alternately through a simple commutator on the axles. As shown in the drawing the batteries were arranged at each end of the carriage. They had a total of forty cells, each with a zinc plate between two iron plates just over one foot (30 cm) square. The plates could be raised out of the wooden troughs containing the electrolyte by a simple windlass arrangement. These batteries proved insufficient and more were added on each side of the carriage, roughly doubling the power. The carriage then ran at about four miles per hour (6·4 km) on level track. Although the contemporary account says the experiments were carried out on the Edinburgh and Glasgow Railway, it does not say the carriage ran from Edinburgh to Glasgow. No indication is given of the actual distance travelled.[17]

Fig. 4.5 *Jacobi's motor with wound armature and stator*
(Photo: Smithsonian Institution)

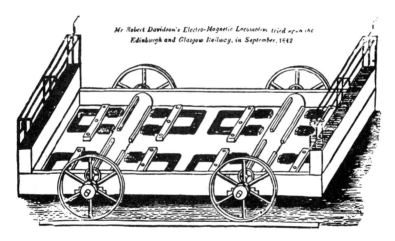

Fig. 4.6 *Davidson's electric locomotive, tested on the Edinburgh and Glasgow Railway in 1842*

An odd feature of the construction was that the cores of the electromagnets were hollow. Each limb of the cores consisted of four iron plates arranged to form a box. According to a contemporary account this construction was adopted to save weight. Clearly Davidson did not appreciate that the total cross-section of iron was important.

It may seem surprising that Davidson was content with simply switching electro-magnets on and off when he must have realised that more power was available if he reversed them. However, magnetic forces were not a limiting factor in Davidson's locomotive. The contemporary account reveals that the power of Davidson's locomotive was limited by mechanical considerations:

> According to Mr Davidson's first arrangement, these magnets were placed so that their poles were nearly in contact with the revolving masses of iron in their transit; but so prodigious was the mutual attraction that the means taken to retain the magnets and iron in their assigned positions were insufficient. They required to be more firmly secured, and their distances had to be somewhat increased, by which considerable power was lost.

Davidson later exhibited his engine and other motors in London. The poster (Fig. 4.7) was printed 'by electro-magnetism', but no other record survives.

4.3   Efficiency and economics

We have enough data to estimate the efficiency of Davidson's machine. The whole locomotive weighed over five tons (5·1 t) and went at four miles (6·4 km) per hour on the level. A horse could pull such a vehicle, and a likely power is about 500 W. The batteries were of a kind capable of giving about 0·2 A per square inch of surface of zinc plate. Each cell had a zinc area of 360 square inches (0·23 m²) and ought therefore to have given about seventy amperes at about one volt. In the original form with forty cells the theoretical power available was therefore about 40 x 70 x 1 = 2800 W. That, however, proved inadequate and the battery capacity was almost doubled. The power available then would have been about five Kilowatts. The overall efficiency of Davidson's locomotive was therefore in the range 500/5000 to 500/3000, or from 10% to 16%.

The efficiency of electromagnetic engines was a matter of some interest in the 1840s. In 1843 Charles Wheatstone described his rheostat, or variable resistance, in a Royal Society paper on Electrical Measurements. The rheostat was developed initially as a measuring device, but Wheatstone stated that it could be used for controlling the speed of a motor, or keeping it constant as the battery varied. He also stated

> Since the consumption of materials in a voltaic battery. . . decreases in the same proportion as the increase of the resistance in the circuit, this method of altering the velocity has an advantage which no other possesses, the effective force is always strictly proportional to the quantity of materials consumed in producing the power, a point which, if further improvements should ever render the electro-magnetic engine an available source of mechanical power, will be of considerable importance.[18]

Fig. 4.7 *Poster advertising an exhibition of Davidson's machines*

This, of course, is not correct since it ignores the power wasted in the resistance, but it suggests that he had conducted some experiments designed to relate the power obtained from an engine and the consumption of materials in the battery.

Economic considerations were heavily against the electric motor. In 1850 the *Philosophical Magazine* noted that

One grain of coal consumed in the furnace of a Cornish engine lifted 143 pounds one foot, whereas one grain of zinc consumed in a battery lifted only eighty pounds. The cost of one hundred weight of coal is under nine pence, the cost of one hundred weight of zinc is above 216 pence, therefore under the most favourable conditions, the magnetic power must be nearly twenty-five times more expensive than steam power. The conclusions being that the attention of engineers and experimentalists should be turned at present, not to contriving more perfect ways for applying electromagnetic power but to the discovery of more effectual means of disengaging the power itself from the conditions in which it exists stored up in nature.[19]

In the year of the great California gold rush, 1849, the United States Commissioner of Patents, Thomas Ewbank, included in his annual report some thoughts on the subject of electric motors.

The belief is a growing one that electricity, in one or more of its manifestations, is ordained to affect the mightiest of revolutions in human affairs . . . When, in addition to what it is now performing as a messenger . . . it can be drawn rapidly from its hiding place, and made to propel land and water chariots . . . then we may begin to think the genius of civilisation is vaulting rapidly toward the zenith.

He referred to the experiments with Jacobi's boat and Davidson's electric locomotive and various other applications of electricity all dependent upon batteries, and then continued in a somewhat pessimistic vein:

but these experiments, interesting as they certainly were, have brought no marked results, nor afforded any high degree of encouragement to proceed. It might be imprudent to assert that electromagnetism can never supersede steam; still, in the present state of electrical science the desideratum is rather to be hoped for than expected. Great, however, will be his glory who in the face of these discouragements succeeds.[20]

The pessimism of the Commissioner of Patents was not shared by Congress. In 1850 it gave $20 000 to Professor Charles Page of Salem, Massachusetts, who had made a number of motors from 1837 onwards. The grant was to develop electromagnetic engines, apparently with the Navy mainly in mind. In a report to the Secretary of the Navy, Page stated that he had made machines of one and four horse power and recommended that he should be granted further funds to build a 100 h.p. motor. He did not get any more money, though in 1854 he built an electric locomotive weighing about twelve tonnes which ran at 19 miles per hour (30·6 km/h) on the level.[21]

4.4 Scientific design

The first electromagnetic engines were designed on a purely empirical basis. Given the basic idea of producing motion by switching an electromagnet which attracts an armature, there are three variables for the designer to consider: the electromagnets may attract either a soft iron armature or a permanent magnet (which could be a continuously energised electromagnet); the electromagnets could either be reversed in polarity or merely switched on or off; and the motion of the armature could be either rotary or reciprocating.

The great problem which designers of electromagnetic engines had to solve was that of obtaining a useful output stroke from a power source which gave great power but through only a short distance. An idea tried by Wheatstone in 1842 was to make the working stroke several times greater than the distance between the armature and the electromagnet. He achieved this by making the armature move at an angle to the line of the magnetic pull.

Three machines incorporating Wheatstone's idea are now in the Science Museum (Fig. 4.8), and will still run if connected to a 12-volt battery. Two of the machines were described by Wheatstone in a patent specification;[22] one of these and the third were described by his friend Daniell in a chemistry text book published in 1843.[23] The first of the eccentric engines was said by Wheatstone to be the least useful of the three but to explain the principle best. It has eight horse-shoe electromagnets supported inside a brass ring 25 cm in diameter. The armature is a soft iron ring of half that size, mounted on a crank with a throw of one centimetre. A static eight-part commutator is fixed to one bearing of the main axis and a single wiper rotates with the shaft. Each of the eight segments of the commutator is connected to one of the horse-shoe magnets; one pole of the supply is connected to the common terminal of all the electromagnets and the other is connected to the wiper through the bearings and frame. The arrangement is such that when the armature is closest to one of the horse-shoe magnets the commutator connects the next magnet in sequence so that the armature rolls around the inner surfaces of the magnets pulling the crank around with it.

The next machine has two fixed horse-shoe magnets inclined at 45° relative to one another. The rotor is a wooden disc about twelve centimetres in diameter carrying four soft iron arcs which are not concentric with the axis. The machine has an eight-part commutator on the shaft similar to the commutator of the preceding machine but alternate segments are connected together and to one of the electromagnets so that in each successive quarter of a revolution first one magnet and then the other acts on one of the four arcs. This machine is not in the patent specification.

The other machine was referred to by Wheatstone as his 'eccentric disc machine'. The armature is a soft iron disc fixed on an axis inclined to the main axis of the machine. One end of the inclined axis is pivoted on the centre line of the machine, and the other end is pivoted on a crank on the main axis. The disc is free to rotate. Four horse-shoe electromagnets are disposed around the edge of the disc on one side of it. As each electromagnet is energised in turn, through the four-part commutator on the main shaft, the disc performs a wobbling motion and the inclined axis describes the surface of a cone, causing the crank and the main axis to rotate.

Front Elevation.

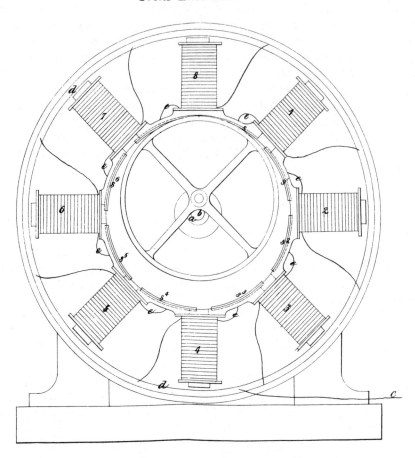

Fig. 4.8 *Wheatstone's 'eccentric electromagnetic engines'*
    a Patent drawing of first design. The inner ring rolls round the inner faces of the
    horseshoe magnets, and also acts as a commutator by pressing on contact pieces *s*

The eccentric disc machine bears a remarkable similarity to an unusual design of steam engine which was developed in the 1830s. This was the disc engine, patented in 1830 by E. and J. Dakeyne and then developed by a number of other people. The disc engine has a barrel-shaped cylinder and its 'piston' is a disc fixed on a ball and socket joint at the centre of the barrel. The disc performs the same wobbling motion as the armature of Wheatstone's eccentric disc machine. Many early electro-

b Machine of first design. There is a conventional commutator at the back

c Machine of second design. The armature surfaces are not concentric with the axis

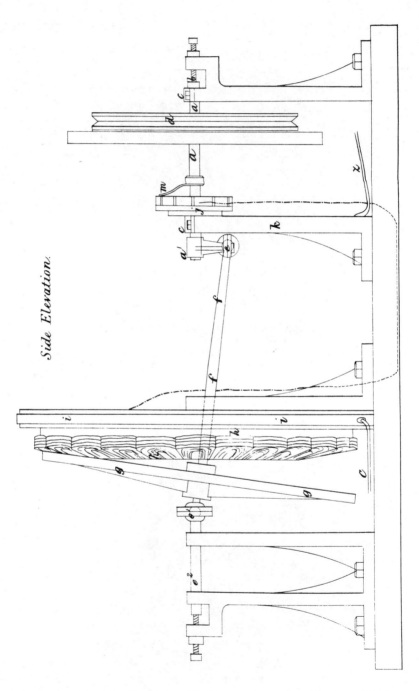

d Patent drawing of the third design. Windings are shown placed in slots milled in the stator iron.

e Machine of third design. The elaborate and modern-looking field windings shown in the patent drawing are replaced by four horseshoe magnets bolted to the frame

magnetic engines were consciously based on steam engine designs, and it is probable that Wheatstone was intending to make an electrical machine analogous to a steam engine whose operating principle had not been tried out electrically.

Only one other person made an eccentric electromagnetic engine, and that was the Frenchman Froment who constructed a number of engines (not all eccentric) between 1844 and 1848.[24]

Most of the people who worked on electric motors during the 1840s thought they were more likely to succeed with a reciprocating machine, using a crank to obtain rotary motion, than with a purely rotary machine. One of these was Uriah Clarke of Leicester, who demonstrated an 'electromagnetic carriage' on a circular railway at the Leicester Exhibition in 1840.[25] Clarke's engine had a single electromagnet which acted on an armature on a pivoted lever (Fig. 4.9). The lever was coupled by a chain to a crank and the idea was that the electromagnet was energised to give the armature, and hence the chain, one brief, sharp pull during each revolution of the crank. (This may sound a crazy idea to modern electrical engineers who expect their motors to have a smooth output torque – but is no worse than the operating cycle of an internal combustion engine!) Clarke's carriage would run for two and a half hours when the battery was charged – but since the total weight was only sixty pounds (27 kg) it must have been fairly small.

Also in 1840 Thomas Wright published in the *Annals of electricity* his idea for increasing the working stroke of an electromagnetic engine.[26] Basically his proposal was that the armature should be hinged at one edge, with that edge being in contact with the electromagnet (Fig. 4.10). In this way he thought he could increase the useful working stroke from ¼ inch to 1¼ inches (6 mm to 30 mm).

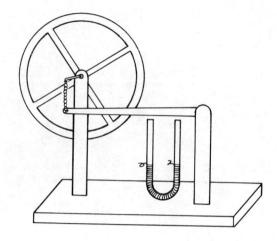

Fig. 4.9   *Uriah Clarke's motor*
The chain gives the crank one brief tug in each revolution

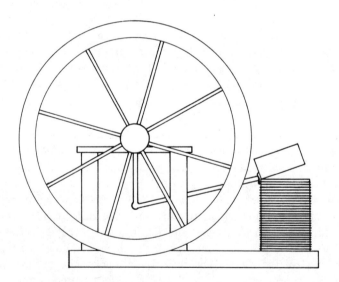

Fig. 4.10   *Thomas Wright's motor*

The most ingenious electromagnetic engine must surely be the machine constructed by Thomas Allan in 1852 (Fig. 4.11). This is basically a reciprocating engine with four cranks and four 'piston rods'. Each piston rod carries four armatures which press on collars on the rod but are not otherwise fixed to it. Sixteen sets of coils, one set for each armature, are energised one at a time by a commutator. Each 'piston rod' is active for one quarter of a revolution of the crankshaft and each electromagnet is energised for one-sixteenth of a revolution. As each armature reaches its electromagnet it is stopped by it, but the piston rod continues its travel.

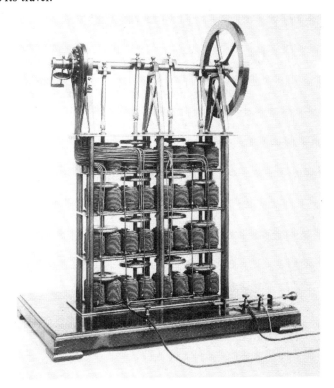

Fig. 4.11 *The Allan engine - the most elaborate of all electromagnetic engines*
Each 'piston rod' is active for one quarter of a cycle, and during that time each of the four armatures on the 'piston rod' is attracted in turn

In the commercial sense all these machines were failures. One reason was their dependence on expensive chemical batteries, but if that were the only reason then the electromagnetic engine would have flourished when generators made electricity readily available in the 1880s. The term 'electromagnetic engine' for these early magnetic engines is unfortunate, but so widespread it can hardly be avoided. In modern terminology these engines were 'magnetic' machines which depended on magnetic attraction between stator and rotor. Modern motors such as the induction motor are 'electromagnetic' machines in which the fields on one side of the air gap generate currents on the other side. Machine theory tells us that magnetic machines get better as they get smaller, while electromagnetic machines get better as they get larger. The future of electric power lay with electromagnetic machines. Du Moncel did not know the theory, but by 1878 he had observed the failings of larger electromagnetic engines:[27]

Attempts have been made . . . to employ the attractive effect of electromagnets . . . as a motive force. This question today still occupies many heads, and, considering the ardour displayed to solve this problem, one would say that it was no less a question than the discovery of the philosopher's stone in mechanics. Without doubt the creation of a non-explosive motor, which would need no one to attend it, which could be situated wherever it would be most convenient without needing a special building, which could be made to act with more or less force according to the work required from it, finally, whose size would not be cumbersome, would be very important . . . but it can almost be predicted that those motors which are most successful when of small size are precisely those which give the worst results when of large size, should they still give any, which does not always happen.

A sad epilogue for a subject on which so much effort and ingenuity had been expended.

4.5 References

1   *Phil. Mag*, 1821, 59, p.241
2   See Chapter 6
3   'Mechanical effect of electricity', *Mech. Mag.*, 1824, 3, pp. 249-250
4   Moll, G.: 'Electromagnetic experiments', *Edinburgh Journal of Science*, 1830, 3, pp 209-218. Also in *Amer. J. Sci.*, 1830, 19, pp.329-337
5   *Phil. Mag.*, 1832, 8, p. 45
6   Dunsheath, Percy: *History of electrical engineering*, Faber, 1962
7   Henry, J.: 'On a reciprocating motion produced by magnetic attraction and repulsion', *Amer. J. Sci.*, 1831, 20, pp. 340-343
8   Sturgeon, William: 'Description of an electromagnetic engine for turning machinery', *Annals of Electricity*, 1836-7, 1, pp.75-78 and plate II
9   Stratingh, S. and Becker, C.: 'Electromagnetische bewegingskracht en aanwending daarvan tot een electromagnetischen wagen', *Algemeene Konst- en Letterbode*, 1835, pp. 402-422
10 Woodcroft, Bennet: *Introduction to abridgements of patent specifications relating to electricity and magnetism, 1766-1857*, 1859, p. 1xx
11 English patent no. 7386 of 1837
12 *Amer. J. Sci.*, 1837, 32, appendix p. 4 and 33, appendix p. 2
13 *Phil. Mag.*, 1835, 7, p. 107, and 1838, 12, p. 190
14 'Taylor's Electromagnetic Engine', *Mech. Mag.*, 9 May 1840, 32, pp.693-696
15 Forbes, P.: 'On the Application of Electro-Magnetism as a Motive Power', *Annals of Electricity*, September 1840, 5, pp.329-240
16 *Phil. Mag.*, 1839, 15, p. 164
17 Mackie, David: 'Prospects of Electro-Magnetism as a Prime Mover, with a Notice of Mr Robert Davidson's Electro-Magnetic Locomotive, lately tried on the Edinburgh and Glasgow Railway', *Practical Mechanic*, 1843, 2

18 Wheatstone, C.: 'An account of several new Instruments and Processes for determining the Constants of a Voltaic Circuit', *Phil. Trans.*, 1843, 133, pp.303-327

19 *Phil. Mag.*, 1850, 35, p. 550

20 Quoted by Frank J. Sprague in *Electrical Engineering*, 1934, 53, pp.695-707

21 MacLaren, Malcolm: *The rise of the electrical industry during the nineteenth century*, Princetown University Press, 1943

22 English patent specification no. 9022 of 1841

23 Bowers, Brian: *Sir Charles Wheatstone*, HMSO and Science Museum, 1975, pp.75-82

24 du Moncel, Th.: 'Les Moteurs Electriques de M. Froment', *La Lumière Electrique*, Paris, 1883, pp. 193-8. Froment's machine of 1847, his 'électro-moteur epicycloidal', which is similar to Wheatstone's, is in the Musée du Conservatoire des Arts et Métiers, Paris

25 *Annals of Electricity*, 1840, 5, pp. 33-34 and plate 1 and pp. 304-305

26 *Annals of Electricity*, 1840, 5, pp. 108-110 and plate 3

27 du Moncel, Th: 'Exposé des applications de l'électricité', Paris, 1878

Chapter 5

Electrical Science

5.1   The mathematics of electricity

One of the people who followed up Oersted's discovery was André Marie Ampère (1775-1836), one of the most distinguished of French scientists. He showed that as the current passed through a battery it had the same effect on a magnetic needle as did the current in the external circuit, so establishing the concept of a complete *circuit*. Ampère made the important observation that two conductors carrying current should have an effect on each other, owing to the interaction of their magnetic fields, and worked out the mathematical relations involved.

Another French physicist, François J.D.Arago (1786-1853) discovered that a current in a coil wound on a piece of iron had the effect of magnetising the iron. Arago was joint founder in 1816, with the chemist Gay Lussac of the *Annales de Chimie et de Physique,* but he is known best for discovering the phenomenon of 'Arago's disc'. He noticed that when a magnetic needle was vibrating its motion was quickly damped if it were close to a metal plate. He made a device in which a horizontal copper disc was rotated beneath a compass needle, and showed that the needle tended to follow the disc, but was not able to explain the phenomenon in terms of eddy currents induced in the disc.

Ohm's Law that the current through a circuit is proportional to the applied e.m.f. and inversely proportional to the resistance was first published in 1827. Georg Simon Ohm (1787-1854) was a Bavarian physicist and mathematician. German was not understood by many scientists in other countries, and Ohm's work remained in obscurity until an English translation sponsored by the British Association was published in 1841. An appreciation of Ohm's law had been important to Wheatstone (who read German) in his early telegraph work, and it is noteworthy that Faraday had been unable to help Cooke in 1837 partly, at least, because Faraday had no mathematical understanding of the electric circuit.[1]

Joseph Henry had been puzzled around 1828 by the fact that, although the electromagnets he was making generally became stronger as the number of turns of wire increased, there came a point where a further increase did not help, or even reduced the strength. Knowing Ohm's law we can explain Henry's problem: the extra turns added to the resistance of the circuit and reduced the current.

The mathematical theory of alternating current circuits emerged during the 1880s and 1890s, with no individual clearly responsible. Hopkinson showed in 1883 that it was important for a.c. generators to have a similar waveform (Chapter 6), and the sine wave seems to have been generally assumed. It is noteworthy that in the 1896 edition of Wormell's *Electricity in the Service of Man*, a textbook of over 900 pages, the mathematics of a.c. are dismissed as 'beyond the scope of the present book'.

5.2   What else can electricity do?

By the middle of the nineteenth century electricity was established as a valuable communicating agent in the telegraph, and as a powerful chemical agent both in electroplating and in the extraction of highly reactive elements. Progress with electric motors, on the other hand, was disappointing. Electric shock treatment was prescribed by some medical men for a variety of conditions, even if the benefit to the patient was dubious. However, instrument makers profited from the market for small hand-driven magnetos, usually provided with a cam-operated contact to break the circuit and give a high voltage inductive pulse (Fig. 5.1).

The potential application of electricity that really seemed worth pursuing was in lighting. Although the glow produced by subjecting a rarefied gas to a high voltage discharge had been observed in the seventeenth century (Chapter 1) it was not

Fig. 5.1   *Typical late Victorian medical electrical machine for giving shocks*

exploited as a practical source of light until nearly 1900. Two possible ways of creating electric light were studied closely throughout the nineteenth century: the carbon arc and the incandescent filament.

The brilliant light produced by an arc between two pieces of carbon was probably discovered by Davy at the Royal Institution about 1802.[2] There is no evidence that he appreciated its practical potential as a source of light, but he made use of the heat of the arc in his chemical work. During the following seventy years many people tried to develop a practical arc lamp, but with little success.

The most important figure in the early history of the arc lamp development is William Edwards Staite. Between 1846 and 1853 he obtained a series of patents relating to electric lighting, including in 1847 a patent for the idea of controlling the regulating mechanism of an arc lamp by a solenoid carrying the lamp current.[3]

Staite gave lectures and demonstrated his lamps in various parts of England. In 1851 he wrote a manuscript history of the electric light, now in the Archives of the Institution of Electrical Engineers, in which he lists more than forty public demonstrations in various English towns.[4] The first was given in the course of a lecture on 'Electric Illumination' on 25 October 1847 before the Sunderland Literary and Philosophical Society.[5] Among his audience was a young chemist, J.W.Swan, who later said that he first thought about developing an electric light after hearing Staite's lecture.[6]

A number of arc lamps which Staite considered to make use of his ideas were exhibited at the Great Exhibition in 1851, without any acknowledgement to him. In May 1851 he wrote a long letter to *The Times* setting out his claim to be the inventor of the 'self regulating system' of arc lamp control. The Editor of *The Times* declined to publish the letter, but Staite included the text in his history of the electric light. Probably Staite did not receive as much credit as he deserved at the time, but any arc lamp was doomed to be an economic failure until a cheaper source of electricity than batteries became available.

Several experimenters must have fused a piece of wire by passing a current through it, and noticed that the wire became incandescent for an instant before breaking. Who first suggested using such an incandescent conductor as a source of light is not known. The problems of finding a conductor that could be maintained at white heat without melting or cracking, supporting it adequately, and preventing oxidation took decades of research. Later, several inventors eventually developed a viable lamp independently, at about the same time.

One early writer to describe an incandescent lamp was William Grove (1811-1896), a chemist who later became a High Court Judge. In 1845 he referred to a lamp he had made 'four or five years ago' and by whose light he had 'read for hours'.[7]

A coil of platinum wire is attached to two copper wires, the lower parts of which, or those most distant from the platinum, are well-varnished; these are fixed erect in a glass of distilled water, and another cylindrical glass closed at the upper end is inverted over them, so that its open mouth rests on the bottom of the former glass; the projecting ends of the copper wires are connected with a voltaic battery. . . and the ignited wire now gives a steady light.

Joseph Wilson Swan (1828-1914) was born in Sunderland, near Newcastle upon Tyne. His formal education ended at 13, but the young Swan took a lively interest in the industries of Tyneside. This was to stand him in good stead in later years. On leaving school he was apprenticed for six years to Messrs Hudson and Osbaldiston, druggists, of Sunderland. Both principals died before the six years were up, leaving Swan free, and he joined his friend John Mawson in his business of chemist and druggist in Newcastle.

Swan appreciated that one of the requirements for an incandescent lamp was a thin carbon conductor that would withstand being made white hot repeatedly. His first experiments were directed towards making a suitable conductor (the term 'filament' came later) by carbonising strips of paper or cardboard.

In 1848 he obtained specimens of different types of paper, cut them into strips, coiled some into spirals, and packed them all in powdered charcoal in a fireclay crucible. They were then baked in a pottery kiln. In some experiments the strips were first soaked in syrup, tar, and other liquids that leave a residue of carbon when heated.

These experiments continued over some years. By about 1855 he had apparently concluded that 'parchmentised' paper gave the best results. (Parchmentised paper is prepared by treating paper with sulphuric acid, which gives a material resembling parchment). Parchmentised paper has a very smooth surface and it remains uniform when carbonised – any non-uniformity leads to 'hot spots' in the filament and limits the operating temperature. Compared with other specimens he tried, carbonised parchmentised paper was solid in texture, elastic and strong.

Swan mounted his filament in a glass jar closed by a rubber stopper through which the connecting wires were passed. Air was removed by an ordinary piston pump. Swan found that he could make his filament glow red hot, but residual air quickly led to oxidation of the carbon.

He needed a better air pump. He also needed a better source of electricity than chemical batteries to make the idea of incandescent electric lighting worth pursuing further. Swan returned to the subject thirty years later, when better vacuum pumps were available and practical generators made electric lighting an economic proposition. During that thirty years he pursued other interests, mainly in chemistry and photography. He produced the first good quality, permanent photographic prints, he worked on the chemistry of photographic emulsions and contributed to the development of photo-engraving. Most significant of all, he invented the silver bromide photographic paper familiar to everyone as the ordinary black-and-white photograph. In 1867 John Mawson, who had become his brother-in-law as well as his colleague, was killed in an accident, and Swan found himself in charge of a substantial chemical manufacturing and retail business. For some years he had no time for speculative research, and he left the subject of electric lighting until 1877 (see Chapter 8).

5.3   Sources of electricity

Before the development of the self-excited generator around 1867 most electricity
came from chemical cells and it was by no means obvious that in future most
electricity would be generated by induction in rotating machinery. Writing in 1852
Edward Highton, a telegraph engineer, observed[8]

> Electricity may be produced in a variety of ways: by friction; by chemical
> action; by magnetic induction; by change of temperature; by the power and at
> the will of certain animals.

He gave similar attention to all these sources, and said that, although only chemical
cells and magnetoelectricity had so far been applied successfully to the telegraph,
the use of thermoelectricity was possible. Highton urged that electric fish should
receive special attention, for if telegraph engineers could understand the working of
the fish when wholly underwater they would probably discover a means of making
submarine telegraphs that needed no insulation!

Thermoelectricity received considerable attention. The basic discovery, that if a
circuit is made of two different metals and the junctions are at different
temperatures then a current flows, was made in 1823 by the German physicist
Thomas Johann Seebeck (1770-1831). The basic arrangement is known as a
'thermocouple', and a group of thermocouples connected together is a 'thermopile'.
The reverse effect, the cooling of a junction when a weak current flows in a circuit
of different metals, was found in 1834 by the French instrument maker Jean
Charles Athanase Peltier (1785-1845). Ohm used thermocouples in his fundamental
study of the relation between e.m.f. and current. Thermoelectricity was attractive
without the need for moving machinery and with no consumption of expensive
materials as in batteries. Wheatstone, Daniell and Henry experimented together
with a thermopile when Henry visited them in London in 1837. However, thermocouples have a low efficiency and have rarely been used in practice, although a
gas-fired thermopile was made in the 1930s for operating a radio.

Batteries such as Cruickshank's (Chapter 1), which were the chief source of
electricity in practice until about 1830, had two important defects: 'local action'
and 'polarisation'. 'Local action' occurs when the zinc plate is not perfectly pure,
which it never is in practice. Electrochemical action occurs between the zinc, the
impurities, and the electrolyte. In 1828 it was found that 'amalgamating' the zinc
plate by rubbing its surface with mercury prevented local action, though the reason
is obscure. Before then the plates had to be lifted out of the battery, or the acid
drained off, when current was not required.

Polarisation was more difficult to cure. When a simple copper-zinc-acid cell is
used, bubbles of hydrogen gas are evolved at the copper plate. The current is
reduced because the layer of bubbles creates a resistance to the current and also
because the cell becomes a hydrogen-zinc cell rather than a copper-zinc one, and
hydrogen-zinc gives a lower e.m.f. The first practical cell which did not polarise

was the Daniell cell, developed in 1836. John Frederic Daniell (1790-1845) was professor of chemistry in King's College London. His idea was the 'two-fluid' principle: instead of having a single vessel of sulphuric acid the zinc plate was immersed in sulphuric acid in one vessel and the copper plate immersed in copper sulphate in another. The two liquids were separated by a porous separator, such as unglazed earthenware. In Daniell's cell no hydrogen was released, but copper was deposited on the copper plate where, of course, it was quite harmless.

Other people made a variety of cells following Daniell's 'two-fluid' principle but using different chemical combinations. Grove's cell of 1839 used a platinum electrode in strong nitric acid and a zinc electrode in dilute sulphuric acid. In this cell hydrogen formed at the Platinum electrode but was quickly oxidised by the strong nitric acid. The Bunsen cell was a Grove cell with the platinum replaced by carbon, which was equally effective but far cheaper.

By far the best known and most widely used of all primary cells is the one developed by Georges Leclanché (1839-1882) in 1868 (Fig. 5.2). This is a carbon-

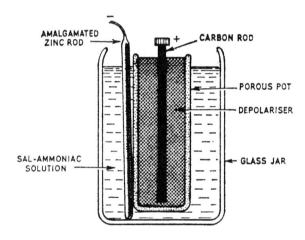

Fig. 5.2 *Leclanché's cell, invented in 1868*

ammonium sulphate-zinc cell with manganese dioxide as a depolariser packed around the carbon. With the electrolyte made into a paste it is the modern dry cell.

Many other chemical combinations have been used. Cells based on mercury and cadmium, such as the Weston standard cell (Fig. 5.3), have two major applications. They are used as voltage standards and also where very small but reliable power sources are required, as in hearing aids.

Secondary batteries, or 'accumulators', which can store electricity and be charged and discharged repeatedly were important for the early supply industry. All d.c. systems had storage batteries to maintain the supply when the generators were not running. The type invariably used was the lead-acid battery, due in its earliest

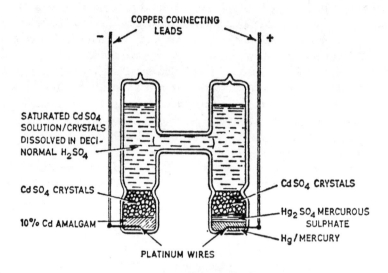

Fig. 5.3 *Weston cell, in the form used as a voltage standard*

form to Planté. Raimond Louis Gaston Planté (1834-1889) made the first practical accumulator in 1859, using two lead plates dipping into dilute sulphuric acid. When a current was passed through the device oxygen was set free at one plate, the anode, and reacted with the lead to form lead peroxide (PbO_2). Hydrogen was set free at the other plate. Planté found that if he disconnected the source of current and joined the plates through a suitable circuit then the device acted as a primary cell. As the cell discharged both plates became coated with lead sulphate. On recharging the coating on the positive plate became lead peroxide again and the lead sulphate on the negative plate became pure lead again. Planté found that the characteristics of the cell improved if the cell were charged and discharged several times, a process known as 'forming'.

Camille Faure, who like Planté was a Frenchman, found a better way of making a lead-acid battery than the slow and expensive forming process. He coated the lead plates with oxides of lead formed into a paste: Pb_3O_4 for the positive plate and PbO for the negative. When the accumulator was charged the PbO on the negative plate was converted into a very spongy lead.

The problem with plates coated with lead oxide pastes was that lumps tended to fall off. Joseph Swan made battery plates in the form of a grid (Fig. 5.4) with the paste pressed into the spaces, and that kind of construction continues to be used although many detailed improvements have been made.

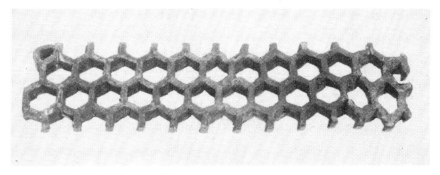

Fig. 5.4   *Grid of Swan's battery plate*

5.4   References

1   Ohm, G.S.: *Die Galvanische Kette Mathematisch Bearbeitet,* Berlin, 1827, and, translated into English, in *Taylor's Scientific Memoirs,* 1841, 2.
2   Journal of the Royal Institution, 22 June 1802
3   English patents nos.   11076 of 7 February 1846
                       11449 of 12 November 1846
                       11783 of 3 July 1847
                       12212 of 12 July 1848
                       12772 of 20 September 1849 and
    British patent no.   631 of 12 March 1853
4   IEE Archives reference SC 52
5   Staite's 'History', p.9
6   See below in this chapter and Chapter 8 for details of Swan's work.
Swan's assertion that Staite's lecture was his starting point is quoted in M.E.Swan and K.R.Swan, *Sir Joseph Wilson Swan FRS,* London, 1929, reprinted Oriel Press, Newcastle on Tyne, 1968. Swan quoted the wrong year (1845), but the facts are established in C.N.Brown: *J.W.Swan and the invention of the incandescent electric lamp,* Science Museum, London, 1978, pp.3-5. (Brown was not aware of Staite's manuscript history, and Staite does not mention Swan).
7   Grove, W.R.: 'On the application of voltaic ignition to lighting mines', *Phil. Mag.,* 1845, **27**, pp.442-446
8   Highton, Edward: *History and progress of the electric telegraph,* London, 1852, p.38

Generators

6.1 Magnetos

Faraday's disc generator of 1831, described in Chapter 2, was the first machine to produce a continuous flow of electricity from mechanical power. Its output, however, was very small because only a single conductor (the radius of the disc) was cutting the magnetic flux. In the following year a Paris instrument maker designed a machine in which all the turns in a coil cut the flux simultaneously. The electromotive force induced was therefore equal to the electromotive force induced in a single conductor multiplied by the number of turns in the coil. This machine was the first machine to combine the basic elements of a modern generator — a coil and a field magnet with one rotating relative to the other.

The instrument maker was Antoine Hippolyte Pixii (1808-1835), son of another instrument maker Nicholas Constant Pixii (1776-1865) who had acquired in 1821 the workshop of his uncles, Dumotiez Frères. The Pixiis were closely associated with the Académie des Sciences in Paris, and Pixii's generator was first announced in a paper read to the Académie by M. Hachette, its Secretary, on 3 September 1832.[1] It is clear from the contemporary account that Pixii's machine, like Faraday's, was seen at that stage as a demonstration of the 'identity of electricities' and not primarily as a source of electric current.

M. Hachette informed the Academy that an apparatus giving a continuous train of electric sparks had been constructed by the son of M. Pixii, the physical instrument maker.

The apparatus is composed of two horse-shoes, with the same opening, one of magnetised steel and the other of soft iron. They are placed end-to-end in a vertical plane, with a common axis of symmetry; the opposing ends approach closely but never touch.

The cross-section of the magnet is a rectangle; that of the soft iron is a circle whose diameter is equal to the length of the rectangle. Many turns of silk-covered copper wire are wound on the two limbs of soft iron . . . The copper wires lead to a glass vessel containing mercury and are fixed so that when the mercury is at rest the ends of the wires are a short distance from the surface of the mercury.

The horse-shoe magnet . . . is rotated on its axis of symmetry and at each half-revolution its poles are close to the ends of the soft iron, which are fixed, and the magnet influences the soft iron over the short distance which separates them. At the same instant the influence is communicated to the wound wire, and produces at the ends of the wire a train of electric sparks which can be seen at the surface of the mercury.

Hachette went on to explain that the magnet was driven through gearing from a driving handle, and that the vessel of mercury was mounted on the same base as the remainder of the apparatus. Consequently, the mercury was vibrated as the apparatus was turned, and ripples on the mercury surface gave an intermittent contact with the copper wires.

Subsequently, Pixii was able to demonstrate the electrical decomposition of water by electricity induced by a rotating magnet. It seems, however, that the apparatus described above was inadequate for that purpose – probably its e.m.f. was too low. Pixii fixed a horse-shoe magnet in his lathe, which could make at least ten revolutions per second, and arranged the soft iron horse-shoe alongside. The ends of the coil wound on the soft iron led to an electrolytic cell.[2]

With a stronger magnet and more turns in the coil Pixii was able to produce electric shocks and to decompose water with the hand-driven machine, but since its output was alternating current the hydrogen and oxygen evolved from the water were mixed together. Ampere had already devised a change-over switch which, because of its action, he called a see-saw ('bascule' in French). According to Ampère himself[3]

M. Hippolyte Pixii had the happy idea of applying to his machine the see-saw which M. Ampère had devised in order to change the current in his electro-dynamic experiments.

Ampere then described how a cam on the machine pushed the see-saw in one direction for half a revolution, and a spring pushed it the other way during the next half revolution. He continued

In the first trial of this arrangement the see-saw dipped alternately into channels full of mercury, as in the see-saws of M. Ampere; but, when the movement became rapid, the mercury splashed out of the channels. M. Hippolyte Pixii avoided that inconvenience by replacing the mercury by little plates of copper. These plates were amalgamated on their surface in order to improve their contact with the points of the see-saw which struck them alternately. By means of this ingenious arrangement, the electric current in the part of the conducting wire which is after the see-saw, always goes in the same direction, from which it follows that it disengages only oxygen in one of the tubes and only hydrogen in the other, and the two gases are obtained separately.

The Académie awarded a prize to Pixii for his machine, a gold medal worth 300 francs. It was one of two prizes in Mechanics founded by M. de Montyon in 1831. The report of the Commission charged with awarding the prizes lays equal stress on Pixii's basic machine and on his use of Ampère's 'see-saw' to rectify the output.[4]

Two complete Pixii machines survive, one in the Deutsches Museum in Munich and one in the Smithsonian Institution in Washington (Fig. 6.1). It is known that one was demonstrated in London, at the Adelaide Gallery in 1833.[5] Apparently a number of machines were made, and the Pixiis had a leaflet printed, entitled 'Instruction pour remonter l'appareil magneto-électrique' and including notes on obtaining sparks and electrolysing water with the machine. There is a copy of this leaflet in the Ronalds' Library of the Institution of Electrical Engineers.[6]

Fig. 6.1   *Pixii generator with 'see-saw' commutator, made about 1832*

Machines similar to Pixii's were soon made by others. In June 1833 Joseph Saxton, an American instrument maker, showed the British Association in Cambridge a magneto of his own design.[7] It has a compound permanent magnet and the ends of the armature winding are brought out to a disc and a double-ended spike on the end of the horizontal shaft carrying the armature (Fig. 6.2). The disc and spike dip in a cup of mercury, and when the machine was turned sparks could be seen as the spike left the mercury and broke the circuit.

Fig. 6.2　*Saxton's magnetoelectric generator, 1833*

The London instrument maker Edward M. Clarke made so many magnetos that he called himself a 'Magnetician'. He and Saxton had an acrimonious exchange in the *Philosophical Magazine* in 1836 in which Clarke was accused of 'piracy' of Saxton's ideas, but in fact both were contributing to the development of the magnetoelectric generator.[8]

The metallic commutator, in place of contacts dipping in mercury, seems to have been the invention of William Sturgeon. He called it his 'unio-directive discharger'.[9]

Emil Stöhrer of Leipzig, another instrument maker, made a series of quite elaborate magneto-electric machines during the 1840s. One now in the Science Museum (Fig. 6.3) has three horse-shoe permanent magnets and an armature assembly carrying six coils. A switch enables the coil connections to be changed so that one, two or three coils are in series. The handle and belt drive makes it possible to rotate the armature at about 500 revolutions per minute, and each coil now generates about 1·5 V, although the magnets have probably weakened over the years.[10]

Fig. 6.3   *Stöhrer's multipole magneto-electric generator*
A switch inside the drum above the coils connects one, two, or three coils in series

The magneto Wheatstone made in 1840 for his first ABC telegraph (Chapter 3) was similar to Clarke's machine. It is now in the Science Museum in London (Fig. 6.4). If the handle is turned at one revolution per second (which is hard work!) it generates 16 V on open circuit. A simple commutator makes the output a uni-directional series of pulses (Fig. 6.5), and all the magnetos so far described would have given similar outputs.

Fig. 6.4   *Wheatstone magneto of 1840*
A commutator (visible between the poles of the magnet) makes the output uni-
directional

In a patent of 1841 Wheatstone described a magnetoelectric machine whose
output is 'not to be distinguished from a perfectly continuous current'.[11] He
thought that such a machine could be used for many purposes for which chemical
batteries were employed, especially to produce those effects which required a

battery consisting of a considerable number of small elements. The new machine, which Wheatstone called a 'magneto-electric battery', had five armatures rotating

Fig. 6.5   *Waveform of the machine in Fig. 6.4*

on a common shaft between six compound horse-shoe magnets (Fig. 6.6). Each armature had its own commutator and the armatures were each displaced relative to the next by one fifth of a revolution. The outputs from the five armatures and commutators were connected in series, and the commutators were made so that the circuit was never broken as the wiper passed from one commutator segment to the next.

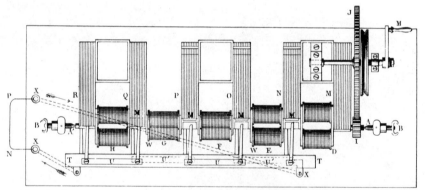

Fig. 6.6   *Wheatstone's 'magneto-electric battery', a compound magneto described in 1841*

The obvious question to ask is why did Wheatstone adopt *five* armatures, not four or six? The specification does say that more or fewer than five armatures could be employed, but five was definitely the preferred number. It is possible that Wheatstone experimented with varying numbers of simple machines linked together. Using a lightly damped galvanometer to observe the output, he could easily have determined which arrangement gave the smoothest output.

If each of the five sections had the same output waveform as the 1840 magneto just described, then the resultant waveform of the 'battery' would be as shown in Fig. 6.7. Four or six of the single waveforms put together would give a less uniform resultant.

No machine of this design is now extant, and possibly no more were made, but in 1873 Du Moncel referred to it as '*the* Wheatstone machine' when discussing early magnetos.[12]

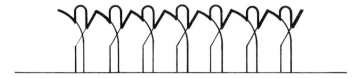

Fig. 6.7   *Waveform of the 'magneto-electric battery'*

In 1845 Wheatstone and Cooke jointly patented the idea that a 'voltaic magnet' (or, as we should say, an electromagnet) should be used in a telegraph magneto in place of the permanent magnet. An electromagnet supplied from a battery could be made much stronger than any permanent magnet at that date, so that the output of such a magneto would be correspondingly greater. But this idea does not seem to have been adopted in practice, and permanent magnet generators remained the norm for another twenty years.

Quite large magnetoelectric generators were constructed for practical use. In 1842 John Stephen Woolrich, a Birmingham chemist, obtained a patent relating to electroplating using a magneto,[13] and his ideas were taken up by the firm of Elkingtons in 1844. The machine Woolrich made then (Fig. 6.8) is now in Birmingham Museum of Science and Industry. It stands 1·6 m high and has four compound horse-shoe magnets supported in a wooden frame. The armature has eight coils arranged around a disc.

Although electroplating provided the first commercial application of electric power, it was soon overtaken in importance by electric lighting. Professor Floris Nollet (died 1853), Professor of Physics at the Ecole Militaire in Brussels, patented a magneto in 1850 which was to be driven by steam power and used to decompose water by electrolysis.[14] The resulting hydrogen and oxygen were then to be used in a limelight apparatus. The Anglo-French Société de l'Alliance, formed in 1852 to develop Nollet's ideas, was probably the world's first company established to produce light by electricity. Nollet soon died, however, and the project was dropped although a number of substantial generators had been made.

An Englishman, Frederick Hale Holmes, who had worked with Nollet was convinced that he could make a generator similar to the Alliance machines and supply an arc lamp. Holmes approached Trinity House for permission to supply generators and arc lamps for use in lighthouses, and Trinity House consulted Faraday. A trial was held at Blackwall in 1857 under Faraday's supervision, and he was delighted with the results. The machine weighed 2 tonnes, stood 1½ metres high, and had 120 coils arranged in 5 rings of 24 on a massive wooden frame. The rotor had 36 compound permanent magnets, each weighing over 20 kg arranged on 6 discs, and required 2½ horse power to drive. The DC output was taken from a commutator by rollers.

Fig. 6.8   *Woolrich magnetoelectric generator for electroplating, made in 1844*
Eight coils revolve between the poles of four permanent magnets, and a commutator
connects two separate plating circuits. (Photo: Birmingham Museum of Science &
Industry)

After this trial and some further development Holmes received an order for two
machines for the South Foreland lighthouse. These machines had iron frames
instead of wood, and the coils were made to rotate while the magnets were static.
They were first used on 8 December 1858. Several more machines of similar though
not identical design were made for other lighthouses. One made for the Souter
Point lighthouse in 1867 was used until 1900 and is now in the Science Museum. It
has 96 coils arranged in eight discs, and gave an AC output (Fig. 6.9).[15]

Magnetos for electroplating and electric lighting all had armatures with a large
number of coils. The electric telegraph required generators that were small and
reliable, but not necessarily powerful, and two basic types emerged. One was the

Fig. 6.9 *Holmes' magnetoelectric generator used at Souter Point Lighthouse from 1867 until 1900*

machine with 'shuttle' armature, developed by Werner Siemens in 1856 and later used in the first Siemens power generators. The other was the induction generator developed by Wheatstone in 1858.[16] In this machine the permanent magnet and the armature coils remain stationary, and pieces of soft iron move so as to vary the magnetic flux through the coils. An alternating e.m.f. is induced in the coils, which was what the ABC telegraphs described in Chapter 3 required. Since there were no commutators or other moving contacts in these machines they proved extremely reliable.

Wheatstone's induction magneto was originally devised for exploding mines, and not for the telegraph. He was a member of the Government's Select Committee on Ordnance, and in 1857 and 1858 he and F.A. Abel, the Chemist to the War Department, investigated the application of electricity from different sources to the explosion of gunpowder. They studied different forms of fuse and three different sources of electricity: the induction coil, Sir William Armstrong's hydro-electric machine, and the magneto. Their conclusion was that for most purposes the best method of detonating gunpowder was to use a fuse designed by Abel fired by Wheatstone's induction magneto (Fig. 6.10) (though they said Armstrong's machine should be used if many fuses were to be fired simultaneously).[17]

Fig. 6.10   *Weatstone's induction magneto of 1858*
Six horseshoe magnets (one removed in the photograph) have a coil on each pole. The coils and magnets are all fixed. On turning the handle a brass cylinder revolves carrying iron bars which pass close to the coils, varying the magnetic flux in the coil. An alternating e.m.f. is thus induced in the coils, which are connected in series

In most induction magnetos the flux through the coils rises and falls in value, but never reverses. Wheatstone made one ingenious machine which actually reverses the flux (Fig. 6.11).

The simple induction magneto gives an assymetric output waveform because the pulse generated as the flux is falling is more 'peaky' than the pulse generated when the flux is rising. There was no direct way of observing the waveform of a current until the end of the nineteenth century. Wheatstone had an experimental magneto (Fig. 6.12) with an arrangement that enabled him to compute the output waveform. An additional section on its commutator has a conductive portion occupying only one thirty-second of the circumference. This additional section can be turned relative to the rotor and the output during one thirty-second of a cycle can be measured with a galvanometer. By running the machine and taking measurements at each of the thirty-two positions the output waveform can be drawn. Fig. 6.13 was derived by the present writer in this way.

For the ABC telegraph Wheatstone sought a square waveform. Before the days of transformers and polyphase systems there was no call for sine waves; alternate

Fig. 6.11   *Induction magneto in which the flux through the single coil is reversed (not merely strengthened and weakened) as a specially shaped iron core revolves*
a   Photograph of actual machine

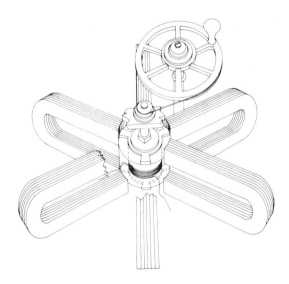

b   Drawing showing principle

positive and negative pulses at peak value were required to drive the step-by-step mechanisms. Wheatstone found that he could obtain a square wave by arranging four coils in his magneto, two on each pole of the magnet, and with the coil centres arranged at the corners of a square. Wheatstone's ABC telegraphs in the 1860s used magnetos with four coils (Fig. 6.14), though his later telegraphs and the machines of other makers used a magneto with shuttle armature.

Fig. 6.12   *Wheatstone's experimental magneto with special contact for determining the output at specific points in a cycle, and hence determining the output waveform*

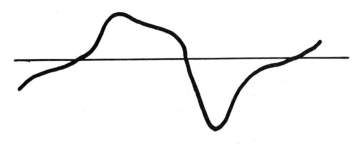

Fig. 6.13   *Waveform of a 'two coil' magneto, derived by using the device shown in Fig. 6.12*

6.2   Self-excited generators

A major step in the evolution of the generator was the introduction of 'self excitation' — the use of a current produced by a generator to energise its own field windings. The concept took a long time to develop, perhaps because it sounded rather improbable. The idea of using the output of a generator to produce the magnetism on which the generator depended must have seemed like lifting oneself up by the shoelaces!

Several people made self-excited generators quite independently in 1866: the Siemens brothers, Samuel Alfred Varley, and Wheatstone, and possibly others share the credit. The Dane Soren Hjorth obtained British patents for generators in 1854 and 1855.[18] Hjorth designed a machine in which an armature was acted upon by a

Fig. 6.14 *Induction magneto with four coils, mounted two on each pole of a permanent magnet, in the base of an 'ABC' telegraph*
This arrangement gives a symmetrical output waveform.

permanent magnet and also by an electromagnet energised by current from the armature. He explained the action of his machine thus:

The permanent magnets acting on the armatures brought in succession between their poles induce a current in the coils of the armatures, which current, after having been caused by the commutator to flow in one direction, passes round the electro-magnets, charging the same and acting on the armatures. By the mutual action between the electro-magnets and the armatures an accelerating force is obtained, which in the result produced electricity greater in quantity and intensity than has heretofore been obtained by any similar means.

Hjorth clearly came very near to making a self-excited generator, but no more is heard of his generators nor the motors he patented at the same time.[19] It seems probable that Hjorth's armature conductors moved past the permanent magnets and the electromagnets in succession, and not simultaneously. Consequently, the electromagnets were not energised at the critical moment when the conductor moved by.

Although Wheatstone had made a wound-field magneto in 1845 there is nothing to suggest that he tried to make a self-excited machine before 1866. As a voracious reader of the scientific literature he probably knew of Hjorth's work, but there seems to have been a general belief that the output of a magneto was limited fundamentally by the strength of the available permanent magnets. This belief (if it was the generally held view) was broken when in April 1866 a paper by Henry Wilde describing some experiments was read to the Royal Society in London by Faraday. Wilde showed that the lifting power of an electromagnet energised by a magneto could be greater than the lifting power of the magnets of the magneto. Wilde's surprise at this result is illustrated by the title he chose for this section of his paper: 'some new and paradoxical phenomena'. In another experiment he used two machines with identical armatures driven at the same speed. One machine had a permanent magnet field and the second had a wound field energised by the output of the first machine. The output of the second machine was eight times that of the first.[20]

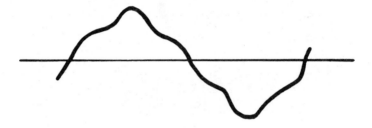

Fig. 6.15   *Waveform of a 'four coil' magneto*

On 14 February 1867 Wheatstone and C.W. Siemens presented similar papers to the Royal Society, both describing self-excited generators. [21,22] Their work had been quite independent, although afterwards they discussed their machines and experiments together.[23] Subsequently, it became known that Varley had applied for a British patent covering the idea of self-excitation, but he abandoned the application after the publication of Wheatstone's and Siemens' work.

In their Royal Society papers Wheatstone and Siemens both explained that an armature rotating between the poles of an electromagnet possessing some residual magnetism will generate electricity, and that if the armature is connected through a commutator to the winding of the electromagnet then the magnet is strengthened, and more current is produced. Thereafter the papers read very differently. Siemens' paper was largely based on the work of his brother Werner in Berlin, and it reads like the work of a practical engineer. The title was 'On the conversion of dynamical into electrical force without the aid of permanent magnetism', and he concluded

It is thus possible to produce mechanically the most powerful electrical . . . effects without the aid of steel magnets, which . . . are open to the practical objection of losing their permanent magnetism in use.

Fig. 6.16 *Pacinotti's machine of about 1860, showing the 'ring' armature*
This predates Gramme's use of the ring armature, though Pacinotti used a solid piece of iron whereas Gramme wound his armature core from iron wire

Fig. 6.17 *Early Gramme generators*
a Laboratory pattern

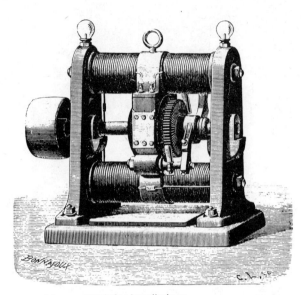

b   The 'A' pattern, used for small lighting installations

c   An 'A' pattern Gramme generator

d Gramme generator of 1884 for large currents

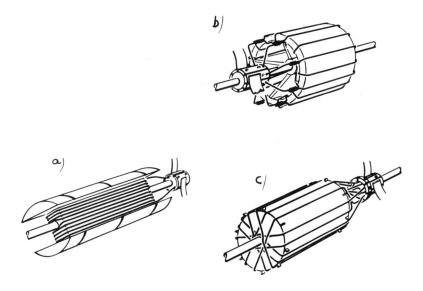

Fig. 6.18 *a* The 'shuttle' armature, so called because the wire is wound on an iron core similar to a weaver's shuttle
    b The ring armature in which the coils are wound around a ring of iron
    c The drum armature in which the whole winding is on or near the outer surface of the iron

Fig. 6.19   *Early Siemens generators*
          a vertical pattern

b horizontal pattern
The horizontal machine has been sectioned, and shows details of the construction
of both the armature and the field

Wheatstone, on the other hand, was writing up an experiment in pure science. His title was 'On the augmentation of the power of a magnet by the reaction thereon of currents induced by the magnet itself', and he set out to show that an electromagnet possessing some residual magnetism could be made into a powerful magnet by currents originated by the action of the magnet itself.

The machine Wheatstone demonstrated on 14 February 1867 survives, complete with its driving arrangements (Fig. 6.20). The stand was furnished with

Fig. 6.20 *Wheatstone's self-excited generator*

ropes, pulleys and handles for two men (I wonder if two students were invited to help with the demonstration). Wheatstone observed that the output of his machine could be increased by placing a shunt resistance across the field winding, though he does not appear to have understood why this helped. After the Royal Society meeting he wrote to Siemens suggesting that he should try shunting the field of his machine, and Siemens found it beneficial. We have a little information about the output of Wheatstone's machine, though it is difficult to convert to modern units. Without the shunt and when driven by two men, it could raise to red heat 10 cm of platinum wire 0·17 mm in diameter. With the shunt it could heat 18 cm of the same wire. Siemens gave no data for the output of his machine, though he did say it could run hot enough for the insulation to catch fire!

Wheatstone's and Siemens' self-excited machines were very similar: both had a long thin shuttle armature and a simple two-pole-field. Several other people were near to making self-excited machines and one, Samuel Alfred Varley, had applied for a British patent covering the idea. Wheatstone was fully occupied with his telegraphs and he did no more work on generators; the Siemens brothers and their British and German Companies pursued generator development vigorously, as did several other firms in Britain, Europe and America.

The self-excited generator made people realise that rotating machines of this type could be made to generate almost unlimited electric power. The invention stimulated the practical development of electric lighting equipment and the first lighting installations all used self-excited generators. Virtually all public supply generators, however, have been separately excited because the system is easier to regulate if the excitation current is separately controlled. In the chain of events leading to public electricity supply, the self-excited generator was not a vital link; but it was a valuable stimulant.

6.3 Practical generators

The first self-excited generators, and others made in the late 1860s, gave an output current that fluctuated considerably. The first really practical generators were made by the Belgian, Zénobe Théophile Gramme (1826-1901), working in Paris around 1870. He was a carpenter who had worked as a model maker for the Société de l'Alliance and became interested in electricity. Gramme's name is always associated with the ring armature, though in fact a machine with a ring armature has been made by the Italian Antonio Pacinotti (1841-1913) about ten years earlier (Fig. 6.16).[24] Pacinotti used a solid iron ring, but Gramme formed his armature core from iron wire. Typical early Gramme generators are shown in Fig. 6.17, and considerable numbers of these machines were sold in the 1870s, mainly for lighting.[25]

The German firm of Siemens & Halske were already manufacturers of small generators for firing mines and for telegraphy. The practical success of Gramme inspired them to design and manufacture powerful and efficient generators for lighting and power transmission. Their designer was Friedrich von Hefner-Alteneck (1845-1904), who had joined the firm as a draughtsman in 1867 and rose to be head of the engineering department. He devised in 1872 the drum armature, which may be regarded as a development of the shuttle armature of the earlier Siemens generators (Fig. 6.18). The new machines were displayed at the Vienna World Fair in 1873.

Surprisingly, Siemens supplied their first generators (Fig. 6.19) to customers seeking mechanical power rather than light. As early as 11 May 1872 Werner Siemens was telling his brother Carl in a letter[27] that 'Hefner's small rotating machine runs just as well as a motor as it does as a generator', but it was still to be several years before he had a customer. In 1877 Krug von Nidda (1810-1885), the Head of the Prussian State Mines, expressed an interest in using electrically driven drills in the mines. Siemens already had mining interests, having bought a copper mine in the Caucasus in 1863. Between 1877 and 1882 the firm carried out an extensive development programme on the applications of electricity to mining. Although steam power had long been used for pumping the mines and for winding engines, the advent of electricity enabled the miner to have power tools at the work face. Von Nidda was also interested in the electrolytic refining of copper, and he purchased equipment for a copper refining plant at Oker, which by later 1878 was producing half a tonne of copper a day.

Siemens' first rock drill was a solenoid operated machine, patented in 1879, and two years later he had a rotary drill also. Siemens' first electric railway locomotive, shown at the Berlin Trade Exhibition in 1879, was designed for use in a mine tunnel, and a complete mine railway was installed in a coal mine in Saxony in 1882. Current was collected at 90 V d.c. from an overhead conductor. One of the original locomotives, which was in use for 45 years, has been in the Siemens Museum in Munich since 1927. Siemens also made a magnetic ore separator in 1880.[28]

The business of Siemens and Halske flourished in Germany, and the formation in 1883 of the Deutsche Edison Gesellschaft, later AEG, stimulated further work for both companies. Werner Siemens welcomed the competition, though there was also considerable co-operation between the firms. They agreed that only the AEG, which was run by the engineer and entrepreneur Emil Rathenau (1838-1915), should build power stations, and that Siemens should supply all the technical equipment except filament lamps. In 1903 after various financial reconstructions the Siemens Company in Germany merged with the electrical business established by Sigmund Schuckert (1846-1895) whose sales organisation seems to have been paticularly effective.

In Britain the Siemens interests under Sir William Siemens had become a limited liability company, Siemens Brothers & Co. Ltd., in 1881. On Sir William's death in 1888, Alexander Siemens (1847-1929) took over the business. He was a nephew of William and Werner whom William adopted, having no children of his own. Under Alexander the British firm concentrated on telecommunications, especially submarine cables. However, in 1899 it was decided to take up power engineering again, and Siemens Brothers Dynamo Works opened at Stafford in 1904. After the First World War the Stafford business became the English Electric Company Ltd.[29]

The first important British manufacturer of generators was R.E.B. Crompton (1845-1940). As a boy and then while serving in the army in India, Crompton was interested in steam traction, and on leaving the army in 1875 he bought a partnership in the agricultural and general engineering firm of T.H.P. Dennis & Co. of the Anchor Iron Works, Chelmsford, Essex. He had intended to develop his transport interests, but circumstances directed his attention to electric lighting instead.

Relatives of Crompton owned the Stanton Ironworks in Derbyshire, and in 1878 he designed a mechanised foundry for casting iron pipes. To be economic the plant had to be in operation day and night, so Crompton investigated the possibility of electric lighting. He visited Gramme in Paris and purchased arc lamps and generators for the foundry. He quickly found that he had become recognised in Britain as an

a   The works and some of its products in 1878

b   The works and some of its products in 1895

Fig. 6.21   *Crompton's 'Fire Medal' struck from scrap metal salvaged after Crompton's works were burnt down.*
The two sides show the works and some of its products in 1878 and 1895.

authority on electrical matters.

Crompton set up an import business, supplying Gramme generators and arc lamps. He became dissatisfied with the arc lamps he was importing and designed an improved lamp which he patented in 1879.[30] In March 1879 he engaged A.P. Lundberg, a Swedish engineer married to an Englishwoman, as his first 'foreman of electrical apparatus shops' to supervise the manufacture of electrical apparatus in a part of the Anchor Iron Works. Crompton offered a starting salary of £3 per week, and made it quite clear that the future prospects were unknown: 'the duration of your appointment will depend entirely on us making your branch of the business pay its way'.[31] In fact the new venture was so successful that Crompton took over the whole of the Anchor Works, renaming it the Arc Works. Within a couple of years he enlarged the works in order to manufacture generators too. From that he expanded into the production of instruments, domestic appliances, and almost everything electrical except cable and filament lamps.

The Arc Works were destroyed by fire in 1895. Crompton presented each employee with a medal struck from metal salvaged from the factory. One side shows the works in 1878, Britain's first electrical factory; the other shows the works in 1895 (Fig. 6.21).

The first generators Crompton made were based on designs by the Swiss engineer Emile Bürgin of Basle. Bürgin had sought to improve on the Gramme armature by replacing the single iron ring of Gramme's machines with a series of rings arranged along the axis. In his original design Bürgin had used square 'rings'; Crompton made them hexagonal (Fig. 6.22 and 6.23). These machines had two advantages over Gramme's: cooling was easier because the armature windings had a greater surface area exposed, and it was much easier to secure the windings adequately against the centrifugal stresses arising in fast running.

Crompton was responsible for many detailed improvements in generator design, including the use of laminated conductors in the armature winding (as well as a laminated core) to reduced eddy current losses in heavy current machines.[32]

It is noteworthy that Silvanus P. Thompson, surveying the evolution of the generator in his classic *Dynamo-electric Machinery*, thought that Crompton and John Hopkinson shared the credit for its successful development − Crompton for his work on the armature and Hopkinson for his on the field.

An improvement patented jointly by Crompton and Gisbert Kapp in 1882 was the 'compound' field winding.[33] In this arrangement some coils of the field winding carry a constant excitation current while others are in series with the armature. The series windings increase the field strength when the load current increases and help to stabilize the output voltage, which would fall with increasing load due to resistive losses in the armature and connecting wires. Equally important, the series field coils compensate for the distortion of the magnetic field caused by current in the armature. In a simple generator the brush position had to be altered at every change of load in order to avoid excessive sparking at the commutator. The compound generator can provide a constant output voltage to supply a varying load without needing constant attention.

John Hopkinson (1849-1898) was Professor of Electrical Engineering in King's College London, and a Consulting Engineer with a special interest in patent matters who was retained to advise Edison in his British interests. He was a keen mountaineer, but was killed with his son and two daughters in a climbing accident in Switzerland.[34]

Fig. 6.22   *Crompton-Bürgin generator*
('Horizontal' version — 'vertical' machines were also made)

The American Thomas Alva Edison (1847-1931) had already established a reputation as an inventor and was living on the proceeds of his inventions in telegraphy when, in 1877, he decided to take up electric lighting.[35] His work on the

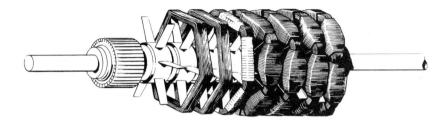

Fig. 6.23 *Details of the Crompton-Bürgin armature*
The iron cores were wound from iron wire on a series of hexagonal formers

filament lamp is discussed in Chapter 8. Edison sought to design and manufacture the complete electrical system, from the generator to the lamp. His early generators were characterised by very long coils on the field magnets, and were nicknamed 'long-legged Mary Ann' (Fig. 6.24). The long magnets were introduced after Francis

Fig. 6.24 *Early Edison generator with his characteristic very long poles*

Upton, Edison's mathematician, recognised that the ampere-turns produced by the field system were an important criterion in generator design. Upton's reasoning was correct as far as it went, but he did not appreciate that part of the magnetic flux leaked between the poles and was wasted. Hopkinson made a number of small models of field systems (Fig. 6.25) and measured the flux which actually reached the armature. He showed that the optimum pole length was much shorter than Edison had been using, and designed the 'Edison-Hopkinson' generator (Fig. 6.26).

Fig. 6.25 *Models made by Hopkinson to study the effectiveness of different shaped field magnets*
He measured the magnetic flux produced at the armature by a given current in the windings

Although d.c. generators run readily in parallel it was found difficult in practice to operate a.c. machines in this way. J.G.H. Gordon who was responsible for a number of a.c. supply systems, said that when two a.c. machines were connected in parallel they 'jumped for three or four minutes'.[36] John Hopkinson showed that a.c. generators would run satisfactorily in parallel, sharing the load, provided they were carefully synchronised first and provided that their outputs had similar waveforms.[37] The practical difficulty is illustrated (Fig. 6.27) by contrasting two generator waveforms shown by Silvanus P. Thompson in the 1900 edition of his book *Polyphase Electric Currents*.[38]

Despite the theoretical analysis it remained the practice for some years to operate a.c. generators singly. The supply network would be divided into groups which could each be supplied by one machine, and at times of light load two or more groups would be switched on to one generator.

Fig. 6.26   *Edison-Hopkinson generator, based on Edison's early generators but modified in the light of Hopkinson's study of the optimum shape of field magnets*

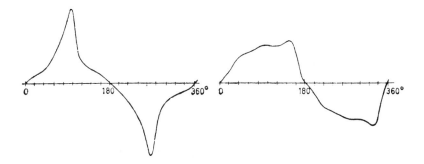

Fig. 6.27   *Output waveforms of two different early generators illustrating the difficulty of operating different machines in parallel*

In his theoretical study of generators Hopkinson evolved the concept of the 'magnetic circuit' of a machine, which could be analysed mathematically in the same way as an electric circuit. The concept was developed independently by Gisbert Kapp (1852-1922). An Austrian by birth, Kapp was first a mechanical engineer but in 1881 he turned to electrical engineering. He became a British subject and married an Englishwoman. From 1882 to 1884 he was employed by Crompton at Chelmsford. Thereafter he worked as a consultant, and then spent a few years in Germany as Secretary of the Verband Deutscher Elektrotechniker and

Editor of the journal *Elektrotechnische Zeitschrift*. In 1905 he became Professor of Electrical Engineering at the University of Birmingham.[39]

Kapp and Crompton had one disagreement which is related in Crompton's *Reminiscences*. Crompton was convinced that he could obtain a much stronger magnetic field in his machines if he used larger field castings and armatures with a much greater cross-section of iron than the Bürgin pattern. He discussed it with Bürgin and resolved to try the experiment, but Kapp was opposed. While Kapp was on holiday Crompton had a machine made, and Kapp returned as it was about to be tested.

> When Kapp on his return found that I had the machine on the test bed, he was rude enough to laugh my ideas to scorn. When the dynamo was started up for the test, it for the first few minutes showed very small electro-motive force at the terminals, and as none of us, Kapp included, then knew that it required considerable time to excite an electro-magnet having a core of great cross-section, Kapp's verdict, 'You have wasted your money, as I told you you would', seemed to be justified. But, just as he spoke, the needle of our voltmeter began to rise, and after a few seconds the machine developed such extraordinary electro-motive force that the belt would not drive the dynamo and flew off. Here ended Kapp's opposition. The first Crompton dynamo was born.[40]

Although early d.c. generators had either drum or ring armatures (if one includes the Bürgin multiple ring), a different pattern was adopted for a.c. machines. The disc machine (Fig. 6.28) had a thin armature containing little or no iron. It had two

Fig. 6.28   *Disc generator for alternating current*
The armature is a series of thin coils arranged around the edge of a disc and passing between the poles of the field magnets. This machine, shown with its exciter, is by Siemens, but Crompton and others made similar machines

great advantages: it was easy to make a multipolar machine and the armature had a low reactance. Multipolar machines were necessary to give a reasonable output frequency when the generator was directly coupled to a reciprocating steam engine. Even the fastest reciprocating engines ran at only about 500 revolutions per minute, requiring a 12-pole generator to give a 50 Hz output. (Output frequencies used in practice ranged from $16\frac{2}{3}$ to 100 Hz). The low reactance made voltage control easier when the load fluctuated.

Disc machines were made in quite large sizes: one rated at 400 kW was installed for public supply at Bristol in 1896.

The disadvantage of the disc generator was that it was impossible to make a three-phase machine of this type. However, the requirement for three-phase generators at the turn of the century coincided with the general adoption of the turbine as prime mover. Most a.c. generators thereafter were directly coupled to turbines and had only one pair, or occasionally two pairs, of field poles and a drum-wound armature (see Chapter 11).

6.4 References

1 Statement (in French) by Hachette to the Académie des Sciences on 3 September 1832, reported in *Annales de Chimie,* Paris 1832, **50**, pp.322-324
2 Further statement by Hachette on 8 October 1832, *Annales de Chimie*, 1832, **51**, p. 72
3 Statement by Ampère (in French) to the Académie des Sciences on 29 October 1832, *Annales de Chimie*, 1832, **51**, p. 76
4 Manuscript report of the Commission charged with awarding the prize in Mechanics founded by M. de Montyon, dated 19 November 1832, in the Archives of the Académie des Sciences, Paris
5 *Literary Gazette*, 1833
6 Ronalds' Library Pamphlets
7 *Annals of Electricity*, 1837, **1**, pp. 145-155
   Trans. & Proc. of the London Electrical Society, 1837-40, **1**, pp. 73-76
   Journal of the Franklin Institute, 1834, **13**, pp.155-156
8 *Phil. Mag.*, 1836, **9**, pp. 262-266 and 1837, **10**, pp. 455-459
9 Dunsheath, Percy: *History of Electrical Engineering*, Faber, 1962 (Dunsheath quotes Sturgeon but does not state the source)
10 Stöhrer, Emil: 'Einige Bemerkungen über die Construction magnetoelektrische Maschinen . . . ', *Poggendorff's Annalen der Physik und Chemie*, 1844, **61**, pp. 417-430 and plate, and also 1846, **69**, pp. 81-93 and plate, and 1849, **77**, pp. 467-493 and plate
11 English patent 9022 of 1841
12 Du Moncel, Th.: *Exposé des Applications de l'Electricité*, Paris 1873, tome 2
13 English patent no. 9431 of 1 August 1842
14 English patent no. 13302 of 24 October 1850 (The patent was granted to Nollet's agent, E.C.Shepard, since a foreigner living abroad could not obtain an English patent)
15 Douglass, Sir James N.: 'The electric light applied to lighthouse illumination', *Min. Proc. ICE*, 1879, **57**, pp. 77-165
16 British patents nos. 1239 and 1241 of 2 June 1858
17 Minutes of the Ordnance Select Committee, in the Public Record Office
18 British patents 2198/1854 and 806/1855

19   British patents 2199/1854 and 807/1855
20   Wilde, H.: 'Experimental researches in magnetism and electricity', *Proc. R. Soc.*, 1866, **14**, pp. 107-111
21   Wheatstone, C.: 'On the augmentation of the power of a magnet by the reaction thereon of currents induced by the magnet itself', *Proc. R. Soc.*, 14 February 1867, **15**, pp.369-372
22   Siemens, C.W.: 'On the conversion of dynamical into electrical force without the aid of permanent magnetism', *Proc. R. Soc.*, 14 February 1867, **15**, pp.367-368
23   This is clear from their subsequent correspondence. For further details see Bowers, Brian: *Sir Charles Wheatstone*, pp. 170-172
24   Pacinotti, Antonio: 'Description of a small electromagnetic machine', *Il Nuovo Cimento*, 1863, **19**, p.383 (Translated into English by S.P.Thompson)
25   Chauvois, L.: *Histoire merveilleuse de Zénobe Gramme*, Blanchard, Paris, 1963
26   Von Weiher, Sigfrid, and Goetzeler, Herbert:*The Siemens Company – its historical role in the progress of electrical engineering*, Siemens AG. Berlin and Munich, 1977, p.35 (Originally published in German as *Weg und Wirken der Siemens-Werke im Fortschritt der Elektrotechnik, 1847-1972*, 1972)
27   Von Weiher and Goetzeler, *op. cit.*, p.36
28   Von Weilher, S.:'Werner Siemens und die Einführung der Elektrotechnik in Bergbau und Hüttenwesen', *Bergfreiheit*, Essen 1953, **18**, pp.388-392
29   Von Weiher and Goetzeler, *op. cit.*, pp. 43 ff
30   British patent no 245 of 1879
31   Letter: Crompton to Lundberg, 4 March 1879, in the Crompton Archives in the Science Museum
32   Thompson, S.P.: *Polyphase electric currents*, 1900, pp. 7-8
33   British patent no. 4810/1882
34   Greig, James: *John Hopkinson*, Sience Museum and HMSO, 1970
35   There are several biographies of Edison, including Josephson, M.: *Edison*, New York, 1959
36   Quoted in Dunsheath, Percy: *History of electrical engineering*, Faber, 1962, p.169
37   Hopkinson, John: 'Some points in electric lighting', *Min. Proc. ICE*, 1883, and continued in *J. Soc. Tel. Eng.*, 1884, **13**, pp. 496 ff.
38   Thompson, *op. cit.*, pp.7-8
39   Tucker, D.G.: *Gisbert Kapp*, University of Birmingham, 1973, and Obituary of Kapp, *Engineering*, 18 August 1922, p. 212
40   Crompton, R.E.B.: *Reminiscences*, Constable, 1928, p.104

Chapter 7

Arc lamps

7.1 The Jablochkoff candle

The development of practical electric generators led to renewed interest in the possibility of electric lighting. The first arc lamps to be used in quantity were the electric 'candles' invented by Paul Jablochkoff (1847-1894) in 1876. Compared with earlier arc lamps (see Chapter 5) and the regulated lamps then available, the candle was cheap and simple. Furthermore, it permitted several lights to be fed in series from one generator.

Paul Jablochkoff was a Russian telegraph engineer. He resigned his post of director of telegraphs between Moscow and Kursk in 1875, intending to travel to Philadelphia to see the exhibition there in 1876. Jablochkoff travelled only as far as Paris, where he met Breguet and was given the use of Breguet's laboratory. Here he actually invented his candle.[1]

The Jablochkoff candle consists of two parallel carbon rods separated by a thin layer of plaster of Paris. The lower ends of the rods are fixed into short brass tubes which secure the candle in a holder and provide the electrical connections. A thin piece of graphite joins the carbon rods at their upper end; when the circuit is switched on the current vaporises the graphite, striking an arc between the ends of the rods. As the carbons burn, the plaster layer crumbles away. For use on a d.c. system, the positive carbon should have twice the cross sectional area of the negative, though in practice most Jablochkoff candles had equal carbons and were used on a.c. circuits. Typically sixteen lamps would be connected in series and supplied from one generator; each lamp would give about 700 candle power and require one horse power, according to a contemporary observer.[2] Another writer gave a figure of 100 candle power.[3] Probably several sizes of Jablochkoff candle were used in practice — the important point was that they were far brighter than gas lights.

The main difficulty with the Jablochkoff candle was that it had to be replaced every time the light had been extinguished, and several inventors made automatic mechanisms for inserting new candles or switching from one to another.

Although the Jablochkoff candle was important in bringing the electric light to public notice, it was soon superseded when reliable, regulated arc lamps became available.

7.2   Arc lamp mechanisms

The requirements for an arc lamp regulator are easily stated: the carbon rods must be brought into contact and then drawn apart when the current is turned on, the spacing must be kept constant as the white hot tips of the rods burn away, the mechanism must be robust, reliable and cheap, and renewal of the carbons must be easy.

All regulated arc lamps have an electromagnet connected in series with the lamp whose function is to pull the carbons apart initially and so strike the arc. Some early regulators used the same electromagnets to control the spacing of the carbons also. As the carbons burnt away and the arc lengthened the current would fall. The Serrin lamp of 1859 used this principle, and when R.E.B.Crompton took up arc lamp manufacture he based his first designs on the Serrin.

In the Serrin lamp (Fig. 7.1) the upper carbon falls under gravity and raises the

Fig. 7.1   *Serrin's arc lamp*

lower carbon at half the speed. While the carbons are moving a star wheel rotates rapidly. When the lamp is first connected to the supply, the carbons are touching and a heavy current flows through the coil of the electromagnet, which pulls down an armature. This armature carries the lower carbon, which moves downwards a few millimetres to start the arc. The armature also moves a detent which comes into engagement with the star wheel so as to prevent the upper carbon falling. As the carbons are burnt away and the current falls, the electromagnet weakens. The detent is pulled back by a spring, releasing the star wheel. The upper carbon can then fall and it moves until the current has increased sufficiently for the electro-magnet to re-engage the detent with the star wheel.

The Serrin lamp was intended for direct current. The upper carbon was made positive, and therefore burnt at twice the speed of the lower, negative carbon. Because the upper carbon moved at twice the speed of the lower, the arc remained in the same position, which is desirable if the lamp is to be used with a shade or reflector.

Crompton placed all the mechanism above the lamp (Fig. 7.2) so that little or no downward shadow was cast, and he lightened the regulating mechanism so that the regulating action took place more often but in smaller stages. A further improve-ment adopted by Crompton was to replace the star wheel and detent by a smooth wheel with a brake pressing on its rim. This control proved so sensitive that the carbons moved only a fraction of a millimetre at a time, whereas previous arc lamps often moved their carbons two or three millimetres. Consequently, Crompton's lamps ran with less flicker than earlier arc lamps.

Arc lamps would not operate satisfactorily in parallel, because the arc itself had a negative resistance characteristic and the internal impedance of the generator provided the ballast necessary to limit the current. The lamp mechanisms considered so far could not be operated in series either, because the regulating electromagnets of all the lamps in circuit would respond to a fall in current, which-ever lamp required regulating. Consequently, every lamp had to be supplied from its own generator.

An alternative way of regulating arc lamps was to monitor the voltage across the arc. As the carbons burn away not only does the current fall but the voltage across the arc increases. Crompton in England and Brush and Wallace-Farmer in the USA found that the regulating function could be performed by a shunt electromagnet of comparatively high resistance (Fig. 7.3). A series electromagnet was still needed to open the carbons a few millimetres and strike the arc when the current was first switched on, but it took no part in the control when the lamp was running. Arc lamps with the shunt regulator could be operated in series from one generator, and this arrangement was widely adopted.

Many different patterns of arc lamp were made in the 1880s and 1890s, but all incorporated the basic principles just described. In some lamps the wheel and brake arrangement was replaced by a clutch acting directly on the carrier of the upper carbon, which was otherwise free to fall, or even on the carbon rod itself. Often the lower carbon was raised by a cord and pulley arrangement as the upper carbon fell,

Fig. 7.2   *Reconstruction of Crompton's first arc lamp*
The electromagnets, which are in series with the arc, depress the lower carbon to start the arc. They also control the fall of the upper carbon by pressing a brake on to a wheel geared to a rack on the rod that carries the upper carbon

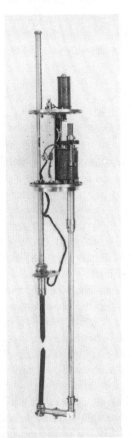

Fig. 7.3   *Probably the earliest surviving Crompton arc lamp*
The series coils are used to depress the lower carbon at starting. The regulating action is controlled by the shunt electromagnet at the top

so that the arc remained in a fixed position.

The best lamp of all was probably the Crompton-Pochin lamp, whose regulator was designed to keep the power in the arc constant by sensing both the current through the arc and the voltage across it.[4]

7.3 Arc lighting installations

In July 1878 *The Electrician* lamented that 'The application of the electric light is in Paris daily extending . . . Yet in London there is not one such light to be seen'.[5]

The Metropolitan Board of Works and the Vestry of Chelsea, then a parish in Middlesex and not yet technically part of London, gave the matter careful thought. The Chelsea Vestry sent their engineer, G.H.Stayton, to Paris with instructions to inspect the electric street lighting there. He reported in August 1878 and his report is worth quoting extensively because it provides both a succinct account of the lighting in Paris and a lucid discussion of the relative merits of gas and electric street lighting.[6]

Gentlemen: In accordance with your instructions I have considered the question of the electric light in lieu of gas for street lighting, and have visited Paris to inspect the system now in operation in that city, with a view to reporting fully thereon.

I had interviews with M. Alphand, Director of Public Works (City Engineer); M. Jabloschkoff, the inventor of the Electric Candle; M. Guichard, C.E., Agent to the Société Générale d'Electricité, and others, all of whom most readily afforded me vauable information upon the subject.

I learn that the Municipality of Paris have contracted with the above Company to light certain streets and places, notably the Avenue de l'Opéra and the Place de l'Opéra, which form a magnificent new street nearly 900 yards in length and 30 yards wide. To effect this the Company have erected 46 lamp columns for the electric lights, at an average distance of 38 yards apart on each side of the street, and have established three electric stations.

The lights consist of the "Jabloschkoff Candle", utilised by means of the "Gramme" Dynamo-Electric Machine, by the aid of which 16 candles (i.e. lamps) are maintained, a steam engine being required to set and keep it in motion.

An electric machine is provided for every 16 lamps, the engine being placed in one instance in the front basement of an unfinished building, in another in a back room on the ground floor of similar premises, and in a third in the basement of the Opera House.

The electric apparatus is worked by a 16 h.p. steam engine with shaft and belts (1 h.p. per lamp being thus absorbed), from which conducting wires for the current are laid in drain pipes under the footways, and carried to the lamps.

Upon a stand fixed on the top of the lamp posts at a height of about 16 feet, opaque globes (Verre Opale) about 18 inches diameter are fixed, and they contain what is called the "Chandelier" which receive the candles.

An electric candle costs 7½d and burns one hour and a half, but the chandeliers are made to receive in advance any given number of candles, and by means of a simple self-acting arrangement when a candle is nearly consumed, the current is passed to a new candle, and the continuance of the light is assured without any visible interruption.

The intensity of a single electric light is stated to be equal to 700 wax candles, but the globes take off about one-third of the light. An ordinary street gas-light in London being equal to twelve or fifteen candles, the superiority of the electric light may be easily understood.

The contract referred to may be termed an experimental one to a certain extent. The company undertook to light the lamps for a period of six months, ending in November next, from dusk till shortly after midnight, and to provide the whole of the apparatus for 1f 45c (1s 2½d) per light per hour. Shortly before the electric light is extinguished about one-third of the gas lamps are lighted and continued till sunrise, the former light being unnecessarily powerful and too expensive to be maintained all night.

The number of lamps provided for lighting the Avenue and Place de l'Opéra by gas is exceptionally great; I should estimate that there are no less than 400 lights. The columns are placed at very short intervals, and have either three or five lamps thereon, consequently the average distance between each light if placed in a line, would be about five yards only.

Although such ample provision is made for gas, the City Engineer says "that the cost of the Electric Light is four times that of gas, but a greater amount of light is obtained." On the other hand, in the lighting of the courtyard of the Hotel du Louvre, it is asserted that a saving of 29$^2/_3$% is effected by replacing 201 gas lamps by 16 electric lights, although 3½ times the amount of light is given.

It is scarcely necessary to remark that the leading thoroughfares in Paris are exceedingly well lighted. There are altogether 39,000 lamps in the city, the annual cost per lamp being £5 16s. Gas is supplied to the Municipality at *one-half* the price charged to private consumers.

The electric light has also been adopted for lighting the Place du Théatre Français, the Madeleine, the Arc de Triomphe, the Orangerie des Tuilleries, the Magasins du Louvre, and about thirteen other places in Paris. It is also in operation in the principal places in Brussels, Madrid and St Petersburg.

Cost of Adoption:- The distance between the lamps in Chelsea being much greater than in Paris, and there being only one lamp upon each column, greatly increases the comparative cost of the systems. They are somewhat irregularly placed, but the distance apart on each side averages about:- 55 yards in Sloane-street, 70 yards in King's-road, 35 yards in Lowndes-square, 35 yards in Cadogan-place, 28 yards on the Chelsea Embankment. In Piccadilly the distance is 30 yards, and in Cromwell-road, South Kensington, 27 yards.

To adopt the electric light for Sloane-street, which is 1,100 yards long and 20 yards wide, would necessitate two electric stations, each of which would require a 16-h.p. steam engine, including shafts, belts, &c, a "Gramme" dynamo-electric machine, a shelter or other building for the machinery and apparatus, the alteration of 16 lamp columns, together with globes, stands, connectors, and the necessary conducting wires, the total expense of which would amount to the sum of £3,200.

The cost of lighting 32 electric candles, including coal, oil, waste, wages, &c, *per hour* would be 16s, and 3,250 hours consumption per annum would be required unless the lights were extinguished and gas substituted at midnight, as in Paris, in which case the annual cost would not be so great.

The present cost of a gas lamp in Chelsea burning 3,850 hours per annum is £3 6s 7d, therefore the expense of the 40 lamps in Sloane-street is 8¼d per hour.

To light the Chelsea Embankment which is about 1,530 yards long, and has 109 gas lamps (including those on the River Wall belonging to the Metropolitan Board of Works), would require a first outlay of £4,800 for 48 lights of 3,250 hours per annum, with an hourly cost of £1 4s. The present cost of the gas lamps is 2s 1½d per hour for 3,850 hours per annum.

To adopt the system in Sloane-square (where there are but 17 lamps) would scarcely be practicable, even if the motive power could be obtained from the pumping engine of the Metropolitan District Railway Station, or from the engine of any manufactory in the locality after the close of the ordinary day's work.

In connection with the foregoing estimates, it must be remembered that in the case of Sloane-street, the amount of light would be 31 times greater than at present, which might be considered as unnecessary expenditure; but the electric current is said to lose 40 per cent of power beyond a radius of about 250 yards, consequently a "Station" for 16 lights has to be established at about every 500 yards which greatly increases the expense. Probably half the above number of electric lights would be found sufficient for effectually lighting Sloane-street, if the quality of the current could be fully maintained at double the distance, by which means alone the cost would be reduced 50 per cent.

GENERAL CONCLUSIONS:- I have arrived at the following conclusions, which may be thus stated briefly, viz:

That the present arrangements for electric lighting are unsuitable for long distances (in this I am supported by the City Engineer of Paris), especially in London, where the lamps are so much farther apart than in Paris. The close proximity of the electric stations is a great drawback to the system, and their establishment in business streets would be a matter of considerable difficulty. These are the disadvantages of the system. The following are the advantages:

About 1½ hours' daily consumption is saved in consequence of instantaneous lighting and extinguishing; the light is vastly superior to gas, and is not injurious; there is an absence of the noxious smells both in the production and combustion; the heat in a room, so often unbearable in the case of gas, is scarcely felt; the most delicate colours are preserved; air is not consumed as in the case of gas;

there is no chance whatever of explosion, and, although the light is so powerful in the streets no accidents to horses have occurred.

Stayton then observed that improvements in electric lighting were 'largely occupying the attention of scientific men'. He concluded that the Vestry ought not to adopt electric lighting at that stage, but he thought there would soon be improvements in the generators and in the distribution arrangements that would make electric lighting worthwhile. He also noted that the economics would improve if 'by some simple method the current could be branched off for household or other requirements'.

On the evening of 14 October 1878 a football match was played under arc lights at Bramall Lane cricket ground, Sheffield. The lighting was arranged by a Sheffield engineer John Tasker (c.1818-1895) who pioneered both electric lighting and the telephone in the city. A portable steam engine behind each goal drove two Siemens generators and each generator supplied an arc lamp mounted on a platform ten metres high at each corner of the pitch. The illuminating power was said to be 8000 candle power, though it is not clear whether that was the power of each lamp or all four. The cost was 1·5 pence per lamp per hour. The match was between two teams formed for the occasion and captained by the brothers J.C. and W.E.Clegg; W.E. Clegg's team won by two goals to nil, and Tasker achieved his object of gaining publicity and obtaining electric lighting business.[7]

Also in October 1878 the Metropolitan Board of Works decided to try electric lighting on the Victoria Embankment. The London agent of the Société Générale d'Electricité offered to contribute the generator and lamps if the Board provided 25 h.p. of motive power and paid for wires and staff. The experiment was agreed, though not without dissent since some members of the Board wanted to see what further technical improvements could be made in electric lighting before committing themselves.[8] About the same time the City of London authorities decided to try electric lighting in front of the Mansion House and on Holborn Viaduct, and in Billingsgate fish market. The market installation was completed first, and was first used on 25 November 1878, when sixteen Jablochkoff Candles provided by the Société Générale d'Electricité lit nearly 4000 m^2. The Embankment was lit from December 1878.

Reference has already been made to R.E.B.Crompton's interest in electric lighting (see Chapter 6). His first lighting sets, consisting of a portable steam engine, generator and arc lamp, were exhibited at shows in 1879 and hired out to public works contractors wishing to work at night and for fêtes and other public entertainments (Fig. 7.4). The Henley Regatta was lit in July 1879, and the grounds and lakes of Alexandra Palace were lit by four Crompton lamps supplied by four Gramme generators (Fig. 7.5). At Christmas 1879 Crompton lit his own house in Porchester Gardens, London, using small arc lamps fed from one of his portable generating sets parked in the Mews at the back of the house.

To encourage prospective customers Crompton published a forty-page book in September 1880. *The Electric Light for Industrial Uses* set out the main facts

Fig. 7.4 *Crompton portable generating plant, for lighting*

Fig. 7.5 *Arc lighting on the lake at Alexandra Palace*

required by anyone considering electric lighting. The book (Fig. 7.6) was widely praised in the technical press for providing this information impartially and in a compact form for the first time. It included a price list of Crompton lamps (about

THE

ELECTRIC LIGHT

FOR

INDUSTRIAL USES.

BY

R. E. CROMPTON,

ELECTRIC LIGHT ENGINEER AND CONTRACTOR.

LONDON: MANSION HOUSE BUILDINGS, E.C.,

AND

ANCHOR IRONWORKS, CHELMSFORD.

PRICE ONE SHILLING.

Fig. 7.6 *Crompton's book on electric light — the first practical handbook on the subject*

£15 each), generators by several makers (from £45 to £100), ancillary electrical equipment and suitable steam engines for driving the generators.

There was great demand for lighting equipment. *The Electrician's Review* of 1880 included the comment that 'The electric lamp that has found most favour with the British public is that designed by Mr. Crompton ... in 1880 the makers have been unable to produce lamps fast enough'. Crompton's lamps were frequently chosen by other lighting contractors. The British Electric Company obtained a contract for lighting St Enoch's railway station in Glasgow. They used six Crompton lamps of 6000 candle power each supplied from six Gramme generators. These replaced 464 gas jets and gave much better lighting for about the same cost. Crompton himself was then invited to provide a similar installation for the North British Railway at the Queen Street station in Glasgow.

Early in 1880 Crompton met J.W.Swan, and the two men co-operated thereafter. Crompton became a director of the Swan United Electric Light Company, and

Crompton & Company manufactured lamp fittings and generators designed for Swan lamps. Many of Crompton's lighting installations included both arc lighting and filament lighting. At Queen Street, Glasgow, he provided Swan lamps for the goods yard as well as arc lights for the main station. Another mixed system was then ordered for a trial in the General Post Office in Glasgow. The sorting room and the telegraph instrument room were better lit and cooler than when 180 gas jets were used, and Crompton was given the contract for a permanent installation.

King's Cross Station, London, was lit by Crompton in 1882, and seems to have been his first installation in which one generator supplied more than one arc lamp. Twelve 4000-candle power lamps were suspended ten metres above the platforms. Four Crompton-Bürgin machines each supplied three lamps, a fifth machine supplied two larger lamps in the station forecourt. Two Gramme generators supplied excitation current and also Swan lamps in the engine room. All the generators were driven by a single steam engine.[9]

7.4   Development and decline of arc lighting

Two important improvements in arc lighting were introduced in the 1890s. The 'enclosed' arc introduced by Marks in 1893 had the arc contained within a small glass tube (Fig. 7.7). This restricted the flow of air and so reduced the consumption

Fig. 7.7   *Marks enclosed arc, which extended the life of the carbons by restricting the air flow*

of carbon by a factor of about five. Thus maintenance costs were reduced, although more power was required.

The second improvement to be introduced in practice, although patented a little earlier,[10] was the addition to the carbons of cores of flame-producing salts. Mixtures of fluorides of magnesium, strontium, barium and calcium were mainly used: they increased the light output and also gave some control over the colour.

In 1890 there were only about seven hundred arc lamps in British streets, and probably a similar number in private use. During the following twenty years about 20 000 were installed, but there was little further growth.[11] The 'half-watt' filament lamp, available from about 1914, had an efficiency equal to that of the arc lamp and required less attention. Few additional arc lamps were installed, although some arc street lights continued in use in London until the 1950s.

7.5   References

1   Wormell, R.: *Electricity in the Service of Man,* Cassell & Co., 1896, p.489
2   Stayton, G.H.: 'Report on the electric light' (Report to the Vestry of the Parish of Chelsea, London), *Electrician,* 24 August 1878, pp.166-167
3   Wormell, R.: *op. cit.,* p.552
4   Bowers, Brian: *R E B Crompton – Pioneer electrical engineer,* HMSO, 1969, pp.9-11, and Wormell, R.: *op. cit.* pp.536-548
5   *Electrician,* 20 July 1878, p.97
6   *Electrician,* 24 August 1878, pp.166-167
7   Farnsworth, Keith: 'The illuminating history of a pioneer called John Tasker', *Quality,* Sheffield, Nov/Dec 1978, pp.35-40, *Sheffield and Rotherham Independent,* 15 October 1878, *Sheffield Daily Telegraph,* 15 October 1878
8   *Electrician,* 26 October 1878, pp.275-276
9   Reference 4 above
10   Bremer's British patent 14704 of 1889
11   For numbers of street arc lamps see Byatt, I.C.R.: *The British Electrical Industry 1875-1914,* Clarendon Press, Oxford, 1979, p.22

Chapter 8

Filament lamps

8.1 A practical filament lamp

The early attempts at making a practical incandescent filament lamp (see Chapter 5) all failed because an adequate vacuum could not be obtained until after the development of the Sprengel mercury pump in the mid-1870s. Several inventors then developed successful lamps independently, including Swan and Edison. It is best to consider the work of the principal inventors separately since, as Sir James Swinburne observed, 'Edison and Swan were hardly racing, as they were on different roads'.[1]

The earliest contemporary record of Swan having an experimental filament lamp is in the *Transactions of the Newcastle Chemical Society* for 19 December 1878. The meeting seems to have been a light-hearted pre-Christmas gathering when several members gave short talks. Swan was Chairman, and made a short contribution himself, in which he

> described an experiment he had recently performed on the production of light, by passing a current of electricity through a slender rod of carbon enclosed in an exhausted globe, the physical results of which experiment he thought might be interesting to the members . . . On passing the current, the rod became heated to such an intense degree as to cause it to glow with great splendour. The glass became coated with a sooty deposit on its inside.[2]

The report does not make clear whether Swan produced the lamp at the meeting or whether he merely described it. An account of the incident given in 1888 (ten years later) implies that Swan did actually produce the lamp at the meeting, and it was already broken.[3] His friend Barnard Simpson Proctor took the lamp after the meeting, broke it open, and analysed the sooty deposit on the glass.[4]

Early in 1879 Swan gave three lectures, all duly reported in the local press, on 17 January at Sunderland Subscription Library, on 3 February to the Newcastle Literary and Philosophical Society, and on 12 March at Gateshead Town Hall (Proceeds in aid of the Children's Dinner Fund).[5] The Newcastle lecture attracted by far the most coverage of the three in the local papers, and there are references to demonstrations of electric light. But there is nothing in the press accounts to prove (or even imply) that Swan showed an *incandescent* lamp. It could have been simply a general interest talk by a local expert. Swan was known to be interested in electric lighting, and at the time Messrs Mawson & Swan were supplying two generators and two 6000-candle power arc lamps to illuminate a local pond for skating.[6]

There is ample evidence that Swan did demonstrate an incandescent lamp at the Newcastle lecture, even though the contemporary press reports do not explicitly say so. Swan himself said so later and several witnesses at the lamp patent trials in 1888 testified that they had seen Swan demonstrate a lamp in Newcastle on 3 February 1879.[7] Probably all three lectures were much the same and a surviving manuscript in Swan's hand is the text he used on all three occasions. It bears a marginal note 'Light up incandescence lamps'.[8] It seems therefore that Swan could not demonstrate a lamp in December 1878, but he probably could in January 1879 and certainly did in February 1879.

The lamp in question (Fig. 8.1) used a thin arc lamp carbon, which Swan obtained from the French manufacturer Carré, for its incandescent conductor. He did not at that stage demonstrate a lamp using his own carbon filaments. As with all early lamps the lead-in wires are platinum, the only metal then available which

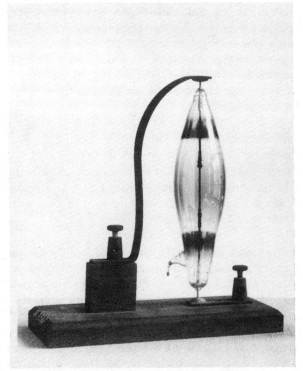

Fig. 8.1 *Swan's first successful filament lamp, made early in 1879*

could be sealed through glass without cracking the glass when heated. According to Proctor, the lamp gave a light of five to ten candle power.[9]

In March 1879 Proctor informed the Newcastle Chemical Society that he had analysed the sooty deposit on the glass, and that it consisted mainly of carbon. At the meeting on 18 December 1879 Swan followed up the subject. He said that the lamp had lasted for about a minute, then the platinum lead-in wires melted. During that minute 'a light was emitted sufficiently brilliant to fully illuminate a room about forty feet square'.

When the globe came to be examined it was found to be coated with a deposit having the appearance of carbon, but as it was difficult to imagine how carbon could have got there, it was desirable that it should be proved whether it was really carbon or not . . . The deposit proved to be for the most part carbon. That was one point settled. Another point which remained to be settled was the question, how the carbon deposit got where it was found? . . . Experiments which I, with the co-operation of my friend, Mr Stearn, have recently made, conclusively prove that the deposit is due to the mechanical transport of carbon particles – the residual air contained within the globe being the medium.[10]

It is clear, from the discussion that followed, that Swan appreciated that residual gas was occluded in the carbon. This residual gas was released when the carbon became hot and was then responsible for the transport of carbon particles to the glass. The process of 'running on the pumps' – continuing to pump the lamp with the filament hot – was not mentioned at the meeting on 18 December 1879 but was the subject of a patent applied for on 2 January 1880.[11]

Less than three weeks later, on 20 January 1880, Swan applied for another patent which described how to prevent fracture of the glass at the seals by using platinum caps.[12] The patent specifications also reveal the progress of Swan's thoughts on filaments. The provisional specification of 20 January 1880 refers to filaments made from cardboard or paper bent or cut into a horse-shoe to prevent breakage from thermal expansion. The complete specification, dated 3 July 1880, refers to cardboard or parchment paper. There is no contemporary evidence of the idea of parchmentised paper being tried before 1880, although Swan's children later said that he had done this in 1855.[13]

One of Swan's technical problems was securing his filament to the lead-in wires. He worked with the stronger Carré arc lamp carbons while investigating the problems of evacuating and sealing the lamp, and then progressed to using carbons of his own manufacture.

Carbonised paper and carbonised parchmentised paper did not entirely satisfy Swan as filament materials. During 1880 he tried other substances and found what he was looking for in parchmentised thread – ordinary cotton thread treated in sulphuric acid. The parchmentising process had the effect of compacting the cotton and making it more uniform, and carbonised parchmentised thread remained Swan's preferred filament material for several years.

Swan demonstrated his lamp publicly for the first time in Newcastle (Fig. 8.2), where he addressed a meeting of the Literary and Philosophical Society on 20 October 1880.[14] A month later on 24 November he gave a demonstration to the Institution of Electrical Engineers in London.[15] He declined to answer questions about his filament material, and a patent covering it was applied for on 27 November 1880. He said one lamp had been running since 8 August – 108 days. If he meant 24 hours a day, that was a life of over 2500 hours, but even if he meant that the lamp was used for an hour or two every day it was unquestionably a practical incandescent filament lamp. Swan's solicitor noted in 1881 that the average life was at least three months when worked four hours a day, or about 400 hours.[16]

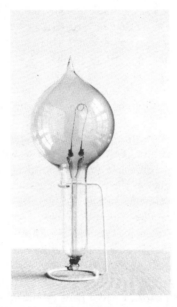

Fig. 8.2 *Earliest commercial pattern of Swan lamp, demonstrated publicly in October 1880*
The long stem kept the heat of the filament away from the point where the wires passed through the glass

Edison became interested in electric lighting towards the end of 1877, after a visit to William Wallace's electrical factory in Connecticut.[17] Wallace made generators and arc lamps, a field new to Edison who reasoned that a commercially successful electric light needed to have similar characteristics to the existing gas lighting. The lamps should have about the same candle power as gas jets and electrical energy should be distributed in such a way that each lamp could be operated independently. Wallace's arc lamps, like most early arc lamps, were operated in series, so that if one failed they all went out. In an interview with the *New York Tribune* of 28 September 1878, Edison said

I have let the other inventors get the start of me in this matter, somewhat, because I have not given much attention to electric lights; but I believe I can catch up to them now. I have an idea that I can make the electric light available for all common uses, and supply it at a trifling cost, compared with that of gas. There is no difficulty about dividing up the electric currents and using small quantities at different points. The trouble is in finding a candle that will give a pleasant light, not too intense, which can be turned on or off as easily as gas. Such a candle cannot be made from carbon points, which waste away and must be readjusted constantly while they do last. Some composition must be discovered which will be luminous when charged with electricity, and that will not waste away. A platinum wire gives a good light when a certain quantity of electricity is passed through it. If the current is made too strong, however, the wire will melt. I want to get something better.

When the reporter suggested that he might easily make a great fortune, Edison replied 'I don't care so much for fortune as I do for getting ahead of the other fellows'.

Edison made optimistic statements about the prospects for electric lighting that excited great public interest, but his first incandescent lamps were complete failures. He experimented with carbonised paper and then in November 1878 stated publicly that he had tried carbon and carbon would not do.

His friend Grosvenor P. Lowrey, a lawyer, had become interested in electric lighting and assisted Edison in the establishment of the Edison Electric Light Company in October 1878, with a capital of $300 000. This money enabled Edison to continue his lighting work.

In October 1878 Edison announced 'I have just solved the problem of the sub-division of the electric light indefinitely'. The announcement received wide publicity and there was an immediate slump in gas shares. The lamp on which Edison based his claim was a platinum-filament lamp including a thermostat to protect the platinum wire from overheating. In fact it was very similar to the lamp made earlier by Maxim, and it proved just as unsatisfactory in practice.

Edison employed a mathematician, Francis Upton, and together they analysed the requirements for a lighting system. They deduced that lamps should have the highest possible resistance so as to minimise the cost of copper in the distribution mains. In the spring of 1879 Edison made high-resistance, platinum lamps with the platinum wound on small spools of clay coated with zirconium oxide. The oxide became conducting when overheated, thus avoiding the need for a separate thermostat, but this lamp did not prove practical either, and in the autumn of 1879 Edison turned again to carbon.

In September and October 1879 he and his team tried out hundreds of different forms of carbon. The lamp usually cited as Edison's first successful lamp was made on 19 October 1879. The conductor was a piece of cotton thread carbonised and bent into a hairpin shape. This lamp is said to have worked for almost two days. However, carbonised cotton was extremely fragile (much more brittle than the carbonised and parchmentised cotton of Swan). He tried paper, and for a public demonstration at the end of the year he used filaments made of Bristol Board, a smooth heavy paper. In 1880 he began commercial production of Bristol Board lamps.

Still looking for better materials, Edison tried a fibre from a piece of bamboo which came to hand and found it gave the best filament so far. He sent expeditions to China, Japan, South America and India to seek other fibrous materials. A particular type of Japanese bamboo was found to be most satisfactory and he arranged for a Japanese farmer to cultivate it specially. The bamboo filament lamp remained the standard Edison lamp in America from mid-1881 until 1894.

Edison quickly sought patents for his lamp in several countries, the British Patent being dated 10 November 1879.[18] Swan did not seek a patent covering his lamp in general terms, though he did patent various points of detail. He felt that it was not patentable because it used only a combination of well known ideas. He must have regretted his omission later.

Fig. 8.3 *Edison's lamp of October 1879*

Swan and Edison were not the only people to develop and market a workable filament lamp. Lane-Fox and Maxim, who, like Swan and Edison, exhibited lamps at the International Electrical Exhibition at Paris in the autumn of 1881, deserve a mention, along with Farmer and the partnership of Sawyer and Man.[19]

Moses Farmer was an American from Massachusetts who had tried making platinum lamps about 1858. He was a teacher who became interested in electricity and made an electric locomotive and a fire alarm telegraph system. He returned to the subject of electric lighting in 1877 and on 25 March 1879 patented a lamp in which the incandescent conductor was a graphite rod held between carbon blocks in an evacuated or nitrogen-filled globe. He proposed parallel operation of lamps, rather than the series circuits normally used then for arc lamps.

William E. Sawyer was a telegraph operator and journalist in Washington, who became interested in lighting. He was short of money until he met Albon Man, a wealthy Brooklyn lawyer, early in 1878. Although Man was supposed to be only the financial partner they carried out their experimental work together. Sawyer and Man discovered the process of 'flashing' or treating carbons by heating them in hydrocarbon vapour so that additional carbon was deposited. They patented the process in January 1879, but they only made and described lamps with a thick rod of relatively low resistance as the incandescent conductor. Most American lamp manufacturers, though not Edison, purchased the right to use the Sawyer and Man patent.

Another American, from Maine, was Hiram S. Maxim (1840-1916). After experience in general engineering and scientific instrument production he also took up electric lighting in 1877. Maxim was one of the three founders and the chief engineer of the United States Electric Lighting Company. This was primarily an arc lighting company, and probably the only one that also tried to develop incandescent lighting. The company acquired Farmer's patents and their first incandescent lamp was a platinum one, working in air. In 1878 they made a graphite lamp.

Maxim introduced two improvements. One, which was not a practical success, was a thermostatically controlled short-circuiting device which permitted the incandescent conductor to cool for a moment when it was getting too hot. His second improvement was the 'flashing' process, which he patented in America in October 1880. Maxim developed this process independently of Sawyer and Man, and his patent described its application to high resistance filaments. The Maxim carbon lamp went into commercial production in the summer of 1880, and installations were made for some years by the United States Electric Lighting Company. Maxim lamps are easily recognised because the filament is held by a bolt at each end, and is usually 'M' shaped. About 1883 Maxim himself turned to other topics and is probably best known for his work on guns and aerial navigation. He became a British subject and was knighted.

St George Lane-Fox was an Englishman. In 1878 he made two types of lamp, neither of which proved satisfactory. One had a high resistance platinum-iridium wire in an atmosphere of nitrogen. The other had a 'burner' of asbestos impregnated with carbon. In 1880, probably inspired by the work of others, Lane-Fox designed a new lamp. He used a fibre of French grass treated with hydrocarbon vapour as Sawyer and Man had done, though Lane-Fox also appears to have developed that process quite independently. Lane-Fox designed a distribution system, using parallel loads and earth return, and designed metering equipment. His patents were acquired by the Anglo-American Brush Electric Light Corporation Ltd, which also acquired Brush's British arc light and generator patents.

When the first international electrical exhibition was held, at Paris in 1881, Edison, Swan, Lane-Fox and Maxim all exhibited their lamps. The Exhibition Jury measured the efficiencies of the various lamps, and found the following results:[20]

	Efficiency of 16 c.p. lamp	Efficiency of 32 c.p. lamp
Edison	196	307
Swan	178	262
Lane-Fox	174	277
Maxim	151	239

The units in the above table are candlepower per horsepower. For those who prefer lumens per watt the conversion factor is $4\pi/746$, or 0.0168, and the table becomes:

	Efficiency of 16 c.p. lamp	Efficiency of 32 c.p. lamp
Edison	3.30	5.17
Swan	3.00	4.41
Lane-Fox	2.93	4.67
Maxim	2.54	4.03

The greater efficiency of Edison's lamp is obvious from the table, but the efficiency and the life of filament lamps are partly under the control of the designer, who may choose a short, efficient life or a long, dull one. Edison understood that, and at the Munich exhibition the following year the lamps Edison exhibited were found to have an efficiency rather less than Swan's.

The Paris electrical exhibition in 1881 embraced every aspect of electrical engineering, but it was most important for its display of lights and lighting systems. Another exhibition was held at the Crystal Palace, in South London, in 1882. Although smaller than the Paris exhibition, the lighting exhibits were better arranged and prospective purchasers could see and compare different makers' products. By showing what could be done these exhibitions, and others like them, encouraged the wider use of electric lighting and helped create a demand for electricity supplies.

8.2   Commercial production of filament lamps

Swan's first commercial lamps were made partly at his house in Newcastle and partly in Birkenhead, at the workshop of Charles Henry Stearn.[21] Stearn was a bank clerk by profession, but no ordinary one. Swan said later that he had met Stearn through an advertisement about Crooke's radiometers, and found a man with the skill and knowledge and equipment to mount a filament in an evacuated glass globe, and a workshop in which to do it. The glass globes were blown for Swan by Fred Topham. The ladies of the Swan household prepared the filaments and Swan carbonised them. Glass blowing, mounting of the filaments, pumping and sealing were carried out at Birkenhead. How the fragile components were conveyed from Newcastle to Birkenhead (about 250 km) is not recorded.

The time had come to establish a company to make the lamps. Swan invited R.E.B.Crompton, from whom he had bought arc lamps although they had not met, to visit him in Newcastle.[22] Crompton was probably present at Swan's lecture in Newcastle on 20 October 1880, and he was certainly present when Swan addressed the Institution of Electrical Engineers on 24 November. He remarked on that occasion 'When I first saw Mr.Swan's beautiful lamp I felt for a moment as if I were a gas shareholder'. On reflection he realised that there was a place for both arc and filament lamps. When Swan's Electric Light Company Ltd was established on 7 February 1881 Crompton was Chief Engineer. The financial backers were all from the Newcastle area. The capital was £100 000, Swan being credited with £25 000 worth of shares. A factory was established at Benwell, near Newcastle, where Fred Topham trained glass blowers to make the bulbs and a staff of young women was recruited to assemble the filaments. The growing electrical industry offered women wider opportunities for employment than they had ever had before. By 1880 telegraph operating was accepted as a respectable occupation, and women were found to be better than men at the delicate work of assembling filaments.

Swan's business expanded rapidly, and a larger company, the Swan United Electric Light Company Ltd, was registered in London on 19 May 1882 with a capital of one million pounds. The initial subscribers included Cuthbert Quilter and Cornelius Cox, both stockbrokers, and Charles Waring, a public works contractor.[23] Crompton was involved in the negotiations leading to the larger company, and in his *Reminiscences* he explained that the demand for lighting required more working capital than the first Swan Company and his own Company could provide.[24]

The United Company published a catalogue early in 1883.[25] It lists the sizes of lamps available – 21 sizes with outputs ranging from 2½ to 100 candlepower and supply voltages from 6 to 100. The price was 5s (25p) except for the three largest sizes which were up to 7s 6d (34½p). The catalogue makes it clear that 20 candlepower lamps were preferred, and these were available in 10 different voltages from 46 to 100. Since current, voltage, and candlepower ratings are given, the efficiencies can be calculated and range from 3·4 to 4·2 lumens per watt.

The Company professed itself willing and able to supply and instal complete lighting plant for 10 to 10 000 lamps, and gave price lists for Siemens and Crompton generators and specimen costings for 50, 100 and 150 lamp installations.

The launch of the Swan United Company was greeted with legal action by the Edison Electric Light Company Ltd, established only two months earlier to exploit the British patent Edison had obtained in 1879. *The Telegraphic Journal and Electrical Review* reported the matter and asked, cynically, 'Is this the beginning of a sham battle preparatory to a combination of the two companies?' Edison himself did not want any merger of interests in which he was not the dominant partner. His attitude was made clear in a letter a year later, on 24 July 1883, in which he told his London lawyer, Theodore Waterhouse, that he would agree to merge his British Interests in a joint company with Swan provided that 'The Company shall be called the Edison Electric Light Company; or at least shall be distinguished by my name in its title solely'.[26] Edison eventually had to agree, however, that Swan's name should also be included in the company title. Swan said he had no objection to the names being put in alphabetical order.

The Edison Company's application for an interim injunction against the Swan Company was heard in the High Court on 28 July 1882. The Plaintiffs submitted an affidavit by John Hopkinson stating that Swan's lamp was virtually identical with Edison's. Swan submitted an affidavit stating that the lamps were quite different. Mr. Justice Chitty refused the application for an injunction pending a full trial.[27]

The case never went to trial, but was withdrawn when the two companies amalgamated. It is interesting to speculate upon what would have happened if the companies had fought on instead of merging. The relevant claim of the patent was

> The combination of a carbon filament within a receiver made entirely of glass, through which the leading wires pass and from which receiver the air is exhausted, for the purposes set forth.

The 'purposes set forth' were, of course, to produce light by the incandescence of an electrically heated conductor. The vital word in that claim was 'filament'.

What is a filament? Was the incandescent conductor in Swan's lamp of early 1879 a filament? Is a rod just over one millimetre in diameter a 'filament' or not? If it is, then Swan's lamp anticipated Edison's claim and the patent was invalid. The point was not fought out in 1882, but it came up again in 1886 and 1888 when the Edison & Swan United Company successfully sued two other lamp makers who were infringing that claim. These cases were eventually settled after long hearings in the Court of Appeal. The Lords Justices decided, by a majority, that whatever the conductor was in Swan's early 1879 lamp it was not a filament.[28] In these proceedings Swan was in the rather invidious position of having to defend Edison's patent and play down his own achievements. We do not know what Swan's feelings were at that time. We do know, however, something of his thoughts in September 1880 before he and Edison came into conflict.

Among Swan's papers is a draft letter to Edison dated 24 September 1880, though the letter is not in the Edison archives and there is no proof that it was actually sent.[29]

T A Edison Esq.   Newcastle
                                                                                        Sept 24/80

Dear Sir,
    For some years I have been working at the problem of Electric Lighting by incandescence of carbon. I have watched with much interest your experiments in

the same direction and the thought has occurred to me that we might with mutual advantage exchange ideas on the subject and to a certain extent share interests; that is to say you might have the benefit of what I have done for America and I have the benefit of what you have done for England and share interests for other countries.

I have made very great progress in some essential points in the construction of lamps. So much so that now I feel quite certain the time has come for undertaking work on a large scale in competition with gas lighting for towns. I think I am in advance of you in several points – especially in the making of the carbons – this I have carried to a very great degree of perfection. I have also ideas with regard to the distribution and measuring of the light such as town lighting would call into operation which I think could be usefully joined with your own.

I can easily convince you if necessary that I have been working long on this subject and that carbonised cardboard was a material that I have for years been experimenting with and was actually working at the very time you announced your use of it. I was also at the time referred to using the simplest possible form of lamp, like your own composed entirely of glass, platinum and carbon, the platinum being fused into the glass and exhausted to a very high degree by a most expert manipulator with the Sprengel pump namely Mr Stearn of Birkenhead.

I therefore had the mortification one fine morning of finding you on my track and in several particulars ahead of me – but now I think I have shot ahead of you and yet I feel that there is almost an infinity of detail to be wrought out in the large application now awaiting development and that your inventive genius as well as my own will find very ample room for exercise in carrying out this gigantic work that awaits execution.

A merger seemed in the best interests of both parties. The Edison & Swan United Electric Light Co. Ltd. was formed on 26 October 1883, to take over all the business of both parties in the United Kingdom. Edison had his name first in the title: Swan did not object to the names being in alphabetical order, even though the Swan Company had a bigger share in the United company than the Edison interests did. The merger made sense: together, and with the patent, they held a monopoly of lamp manufacture in this country.

The lamps made by the Edison and Swan United Electric Light Company were 'Swan' lamps with filaments of carbonised parchmentised cotton. Swan himself made one further major contribution to lamp development: the 'squirted' filament.

The parchmentising process only resulted in a uniform filament if the original cotton fibres were reasonably uniform. Swan wanted a fibre that would be even more uniform than any found in nature, and he applied his chemical experiences and knowledge to the problem. Cotton is basically cellulose, and cellulose reacts with nitric acid to give nitrocellulose, a substance soluble in acetic acid. Swan found that if he squirted nitrocellulose solution into a suitable coagulant, such as alcohol, he obtained reconstituted cellulose. Under the correct, closely controlled, conditions the reconstituted cellulose could be obtained as a thread finer, longer, and far more uniform than any fibre previously available. Swan's patent for the 'squirting' process was dated 31 December 1883, but it was not introduced commercially for several years.[30]

The problems of introducing the process on a commercial scale were considerable, but also the United Company was in a strong monopoly position and the business men running it had little incentive to make drastic changes in their manufacturing processes. In 1884 the Company moved to a new factory at Ponders End, just north of London, but the new factory would have been designed and equipped before the squirted filament process had developed to the point of being commercially practical.[31] Similarly in America the Edison Company continued for some years the manufacture of bamboo filament lamps.

The squirted filament was manufactured in Britain from about 1886 by the Anglo-American Brush Electric Lighting Corporation, one of the companies sued for patent infringement by the Edison & Swan Company, as mentioned above. Anglo-American Brush could have appealed to the House of Lords, but the two companies reached an agreement in July 1889 under which Anglo-American Brush gave up filament lamp manufacture, and Edison & Swan agreed to pay a royalty under some of Lane-Fox's patents, then owned by Anglo-American Brush.[32]

By 1890 the Edison & Swan Company was offering a wide variety of lamps.[33] Standard candlepowers quoted were 1, 2½, 5, 8, 16, 25, 32, 50, 100, 200, 300, 500 and 1000. Standard voltages quoted were 3 to 105V, and to special order lamps could be obtained for up to 160 V. Lamps up to 16-candlepower (the most commonly used – roughly equivalent to a modern 25-W tungsten filament lamp) cost 3s 9d (19p), 50-candlepower lamps (about equivalent to 60-W tungsten filament lamps) cost 5s 0d (25p), and 1000-candlepower lamps (equivalent to rather more than 1000-W tungsten filament lamps) cost 30s 0d (£1-50). Lamps could be supplied with obscured or colour-varnished glass or even made from coloured glass at an extra charge. The large, higher rated lamps were fitted with a surrounding netting of fine wire to reduce the risk from falling glass if a lamp were accidentally broken. Every lamp was marked with two figures, the rated voltage followed by the candlepower, and the efficiency was stated to be 3½ W per candlepower, or 3·6 lumens per watt.

When the basic Edison lamp patent in Britain expired in 1893, several new manufacturers sprang into existence and lamps began to be imported from Europe. While the 16-candlepower lamp had been sold in Britain for 3s 9d (19p), similar lamps had been 1s 0d (5p) in Germany and 2s 0d (10p) in the USA.[34]

One of the new manufacturers was the Dutch company, Philips Gloeilampen-fabrieken. Having begun with filament lamps they took up radio valve manufacture after the First World War and now make virtually all kinds of domestic electrical appliances.

8.3 Lighting installations

When Swan spoke at the Newcastle Literary and Philosophical Society in October 1880, his friend Sir William Armstrong (later Lord Armstrong, the industrialist) was in the chair. Shortly afterwards Armstrong's house, Cragside, Rothbury, was lit by Swan lamps supplied by a waterwheel and generator already on the Cragside estate. Swan described the installation thus

> So far as I know Cragside was the first house in England properly fitted with my electric lamps. I had greatly wished that it should be and when I told Lord Armstrong so, he readily assented. There had, previously to the introduction of the incandescent lamps into the house, been an arc lamp in the picture gallery — that was taken down and my lamps were substituted. It was a delightful experience for both of us when the gallery was first lit up. The speed of the Dynamo had not been quite rightly adjusted to produce the strength of current in the lamps that they required. The speed was too great and the current too strong. Consequently the lamps were far above their normal brightness, but the effect was splendid and never to be forgotten.[35]

After the initial trial in the picture gallery Armstrong had the installation extended and by mid January 1881 there were 45 lamps connected, each of 25 candlepower. Armstrong said that he never used more than 37 lamps at a time, so the maximum output was 925 candlepower. The generator absorbed six horsepower. It was a Siemens machine originally installed for supplying the arc lamp in the picture gallery, and the current was transmitted along a double line of copper wire nearly two kilometres long, supported on telegraph poles.[36]

Armstrong published a detailed description of the installation in a letter to *The Engineer* in January 1881.[37] His comments on the difficulty of controlling the voltage so that the lamps were not overrun and burnt out remind us that no one had yet solved the problem of automatically regulating the output of a generator when supplying a variable load. Armstrong observed that fortunately he had no fuel bill.

> It is important to the preservation of the lamps that the amount of motive power applied should always be proportioned to the number of lamps in light at one time. In my case, I escape the necessity of varying the motive power by using a resistance coil to represent the resistance of each section of lamps which it is desirable to have the option of throwing out of use. By means of these coils the number of lamps in light at one time may be greatly varied without affecting the work of the generator, because the resistance to the current is the same, whether it passes through the coils or the lamps. This method is wasteful of power, but I can afford to waste that which costs me nothing, and is always

sufficient in quantity. If steam or gas engines were employed the case would be different, and there might be difficulty in effecting such momentary adjustments of power as would save the lamps from disturbance, where the number in use was liable to great and sudden variation.

The Swan United Electric Light Company's catalogue of 1883 lists more than one hundred houses and other buildings in which Swan lamps had been installed. Some had been installed by the Swan Company itself, some by Siemens Brothers, Crompton, and others. Another flourishing market was for lighting on board ships, and the same catalogue lists 25 vessels making use of Swan lamps. The safety of electric lighting, compared with oil lamps and candles, was attractive to ship owners, and it was easy to drive a generator from the ship's engines.

A prestigious contract won by Crompton in 1882 was for providing a mixture of Swan lamps and arc lamps in the new Law Courts, then being built in the Strand (Fig.8.4).

The success of these installations brought Crompton many other contracts.[38] One was for lighting Berechurch Hall, near Colchester, being built for the brewer Octavius Coope. Crompton later wrote of it:

This was the first time that a house and premises had been lighted throughout with electric light. It was the occasion, moreover, of a novel departure in handicraft. In order to deal with the novel work of wiring and fixing, I took over from the builder of the house a party of his workmen whose previous trade had been that of bell-hangers. Several of these men gradually bettered themselves, and eventually went into business as master contractors for electrical wiring of houses.[39]

Coope wrote a long letter to *The Times* about the lighting of his house

The light is quite as easy to manage as gas, while the softness, the purity, and the agreeableness are such that a return to any other method of illumination would be now quite out of the question. The pictures, books, and decorations have no chance of injury; the ceilings and walls remain unspoiled, while the difference in health felt after sitting for an evening in a room electrically illuminated, and another lighted by gas, must be experienced before it can be appreciated.

I will now give details as to cost: Estimate for gas plant, buildings, and erection, £740; gas main to house, £75; laying pipes in house, £200; cutting and making good again, £50; chandeliers and brackets, &c., £268 18s. − total, £1,333 18s. Cost of electric light: Four dynamo machines, £405; 220 Swan lamps, £55; 200 sockets for same, £10; cable, wires, switches, and cut outs, £66 4s; cutting and making good walls and floors and incidental work, £60; engine and boiler, extra flywheel and belting, £300 6s; counter-shafting, £25; foundations to engine and flooring, £40; erection, laying wires, &c., and carriage, £90; buildings, £150; chandeliers and brackets, &c., £268 18s. Total, £1 470 8s.

THE ELECTRICIAN, DECEMBER 16, 1882. 113

THE ELECTRIC LIGHT AT THE NEW LAW COURTS.

We this week illustrate the interior, as seen on the day of the formal opening by the Queen, of the new Law Courts. Our illustrations show the large hall lighted by six arc lamps, but these are to be replaced by clusters of incandescent lamps.

Fig. 8.4 *Electric light at the new Law Courts, 1882*
The large lights are Crompton arcs, the small ones Swan filament lamps

The cost of the first outlay for electricity is, as will be seen, somewhat in excess of the same for gas; but then I have no nuisance of lime, or of tar, or other refuse products; no leakage of gas into the house, no smell in the manufacture, or damage to my garden; and, in the place of an unsightly gasometer, I have a compact little engine, placed out of view, and which, when not driving the dynamos, is utilised for pumping water to the top of the house, and can also be employed for sawing wood, or any other purpose of a like nature I may require.

I will now estimate the annual working expenses, and, in making a comparison, I will assume that it would cost the same to manufacture 1000 cubic feet of gas at Berechurch as it does here. [His other house at Brentwood.]

Electric light − 200 18-candle lamps, each working 1150 hours per annum. − Coal at 20s; per ton, £38 10s 1d; engine driver at 30s per week, £78; renewal of lamps, 153 at 5s each, £38 5s; depreciation 10 per cent. on cost of machinery, £74; depreciation 5 per cent. on conductors, £4. Total £232 15s 1d.

Now for less than half this amount of light it costs me £200 a year here, and, therefore, I am, if anything, understating the expense when I say that, had I used gas at Berechurch giving an equivalent illuminating power, it would have cost me at least £400 per annum to produce.

This would show an annual saving of £167 4s 11d to me by using electricity. Having had 20 years experience of lighting this house by gas − which I consider a great improvement upon any previously known method − I am only too sensible of its drawbacks, and, although I am not doing away with it here yet, I am well satisfied that I have adopted electricity at Berechurch.

Faithfully yours, Octavius E. Coope

It may come as a surprise to the modern reader that the country house owner of the 1880s who wanted something better than oil lamps had first to choose between building a gasworks and building a generating station. For this house Crompton installed four 60 V Bürgin generators driven by a single steam engine to supply the 59 V Swan lamps.

Sir William Thomson (later Lord Kelvin) lit his own house near Glasgow. The power source was a gas engine which drove a Siemens generator capable of supplying 50 lamps. A lead-acid battery of 120 cells (3 groups of 40) maintained the supply when the machines were not running. In February 1882 Thomson described the installation in a letter to W.H.Preece, who was also lighting his house in Wimbledon.[40]

Before I left Glasgow on Saturday I had my installations so far complete as to have ample light for every part of the house, from the attics to kitchen, including every lobby and closet − even a trunk room under the slates and boot-cleaning closet in the lower regions, lighted by Swan lamps. I had dining room and drawing room lighted for two or three evenings when we had friends with us and the result was greatly admired. The high incandescence required for good economy is too dazzling and I believe would be injurious to the eyes if un-mitigated. I have found that very fine silk paper round the globe spreads out the light quite sufficiently to make it perfectly comfortable to the eye while con-suming but a small percentage of the light and Lady Thomson has accordingly made little silk-paper globes, already for nearly all our lights (112 in all), some of them in the drawing room slightly coloured red and yellow and green and blue gray, all very faint, which looks very pretty when lighted at night. I have some Edison lamps mounted also and am only waiting to get convenient resistances made and adjusted to keep them in permanent use, the potential I use (90 volts) not suiting them.

Crompton spent much of the years 1885 to 1889 in Vienna lighting the Opera House, theatres and public buildings. The Ring Theatre, previously lit by gas, was destroyed by fire in 1883 with great loss of life. Emperor Franz Josef decided that, for greater safety, electric lighting should be introduced. The Imperial and Continental Gas Company, who retained the contract for lighting the theatres, engaged Crompton to design and supervise the work and to manufacture some of the equipment.

A single central generating station supplied all the buildings, which were up to one mile away – a far more widespread system than any previously attempted anywhere in the world. Although this was still a 'private' supply system the operating conditions were approaching those that would arise with a system of public electricity supply.

The highest voltage filament lamps then available required 100 V. To obtain the advantages of high-voltage distribution (lower current and therefore cheaper cables) he devised the 'five-wire system' of supply with batteries at each load point. The generating station operated at about 400 V d.c. and charged a Crompton-Howell battery of 200 cells at each theatre or other consumer. The theatre lighting was divided between four circuits each supplied at 100 V from one quarter of the battery. In the generating station six Willans 150 horsepower steam engines were each directly coupled to a Crompton generator. The generating capacity was over 700 kW, and for special effects at the Opera 1000 kW could be utilised, with the batteries discharging at their maximum rate.

The Opera House was the first large theatre to be lighted electrically. Bracket lamps were put on the fronts of the balconies and boxes, which was impossible with gas lighting because of the rising fumes and hot air. On the stage Crompton intro-duced coloured lights, separately controlled by resistance dimmers in which resistance coils were short-circuited by being lowered into a vessel of mercury. Several operas and ballets were put on to test the scenic possibilities, and Viennese society flocked to see what could be done with the 'electric gas'.[41]

The experience gained in early private lighting installations like those described above encouraged engineers and others to promote public electricity supply schemes. It was to be many years, however, before public supply could reach isolated country houses, and private generating equipment continued to be installed in many such establishments. An example of such an installation is Chatsworth House, Derbyshire, the home of the Dukes of Devonshire, where electric light was installed in 1893 by Bernard Drake, of Drake & Gorham. The work was completed by November of that year.

The power source was a nine acre lake 120 m higher than the house. A pipe had already been laid from the lake to a point near the house to supply water to a fountain, and Drake found that by tapping the pipe he could obtain 150 horsepower, apparently without spoiling the fountain. This was adequate for supplying the proposed installation of 850 lamps. Since the most common lamp size at the time was 16 candle power, and the efficiency of carbon filament lamps then was about 300 candlepower per horsepower, the maximum electrical power required would be 45 horsepower, or 33 kilowatts, ignoring losses. Drake installed turbines and generators totalling 125˙ horsepower. The generator installation is shown in Fig. 8.5.

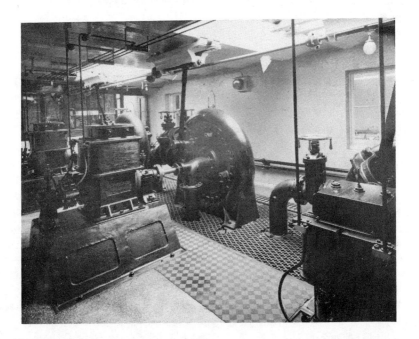

Fig. 8.5 *Electric lighting plant at Chatsworth*
These photographs from the Chatsworth archives were probably taken in 1912.
a The water turbines and Siemens generators. The sets in the middle and at the back were rated at 50 horsepower; the set on the right was rated at 25 horsepower. The back of the switchboard can be seen at the back, left

Three Vortex turbines by Gilkes of Kendal, two rated at 50 horsepower and one at 25 horsepower drove directly-coupled generators by Siemens Brothers, London. Two machines were rated at 300 A, 115 V at 1000 r.p.m. The third machine was rated at 90 A, 150 V at 1250 r.p.m. All three are series-wound non-compound machines, with copper gauze brushes pressurised by steel springs. According to a contemporary description in the *Electrical Review*, the house was supplied normally

b The switchboard controlling the generators. The large hand-wheels operate valves in the water supply to the turbines

by one or both of the larger machines, and the smaller machine charged batteries in an adjoining room. The battery room had 60 cells (presumably lead-acid, though this is not stated) rated at 120 A for four hours. The voltage at the house was 105, and in emergency the batteries and the small generator in parallel could supply the house.[42]

The system was modified in 1912 by the addition of an 'autobooster'. This was a motor-generator unit which generated up to 35 V and 180 A to boost and stabilise the supply to the house. Probably tungsten filament lamps were installed in 1912, in place of carbon ones, and tungsten lamps require better supply regulation. The modifications were supervised by A.W.Sclater (1865-1921), an engineer who specialised in the lighting of large country houses.[43]

The provision of private generating plant for country houses was a flourishing business until after the Second World War. Fig. 8.6 shows an early house lighting plant reconstructed at the Engineerium in Brighton. Fig. 8.7 shows how gas brackets were sometimes 'converted' to electricity.

Fig. 8.6   *Reconstruction of an early house lighting plant*
(Photo: Brighton Engineerium)

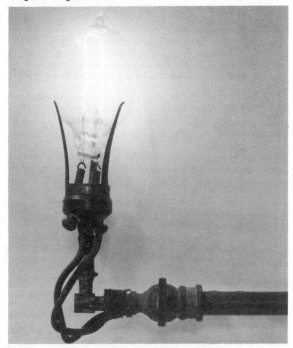

Fig. 8.7   *Swan lamp in a holder designed to screw into a gas bracket*
The wire would normally be draped more elegantly than in this photograph

8.4 References

1 Swinburne, James: 'Sir Joseph Swan's Electrical Work', *J. IEE*, 1931, 67, pp.291-2
2 Report of meeting on 19 December 1878 in *Transactions of the Newcastle Chemical Society*, 1877-80, 4, p.190
3 Brown, C.N.: *J W Swan and the invention of the incandescent electric lamp*, Science Museum, London, 1978, pp.11-17
4 Report of meeting on 27 March 1879 in Reference 2, p.292
5 *Sunderland Daily Echo*, 18 January 1879, p.3
   Newcastle Daily Chronicle, 4 February 1879 p.4 and 13 March 1879, p.3
6 *Newcastle Daily Chronicle*, 29, 30, 31 January and 1 February 1879
7 Reference 3 above
8 Manuscript in Tyne & Wear Archives, Swan Papers
9 Reference 4 above
10 Report of meeting on 18 December 1879 in *Trans. Newcastle Chem. Soc.*, 1877-80, 4, p.327
11 British patent no. 18 of 1880
12 British patent no. 250 of 1880
13 Swan, K.R., and Swan, M.E.: *Sir Joseph Wilson Swan FRS*, London 1929 and Oriel Press, Newcastle upon Tyne, 1968
14 *Newcastle Daily Chronicle*, 21 October 1880, p.3
   Electrician, 1880, 5, pp.279-282
15 *J. Soc. Tel. Eng.*, 1880, 9, pp.339-362
16 Manuscript in Tyne & Wear Archives, Swan Papers
17 There are several biographies of Edison, including
   Josephson, M.: *Edison*, New York, 1959
   A detailed, near-contemporary account of Edison's work is found in Pope, F.L.: *Evolution of the incandescent electric lamp*, Elizabeth, New Jersey, 1889
18 British patent no. 4576 of 1879
19 A good survey of the work of several filament lamp pioneers in Europe and America is given in Bright, Arthur A.: *The electric lamp industry*, Macmillan, New York, 1949
20 Figures taken from the official report of the Exhibition
21 Information about Swan and his Companies has been derived from the biography by his children (reference 13 above), Company records in the Public Record Office and financial notes in *Electrician* and *Electrical Review* of the time
22 This incident is also recorded in Crompton, R.E.B.: *Reminiscences*, Constable, 1928, pp.93-96
23 Company file in Public Record Office; Company No 16841
24 Crompton, Reference 22 above, p.107
25 *Catalogue* of the Swan United Electric Light Company dated 1883 (The catalogue quotes a newspaper report dated 30 November 1882)
26 Letter in the Edison Archives, West Orange, New Jersey
27 *Report of Patent, Design and Trade Mark Cases*, 1882
28 *Report of Patent, Design and Trade Mark Cases*, 1886-1889
29 Manuscript in Tyne & Wear Archives, Swan Papers
30 British patent no. 5978/1883
31 *Electrician*, 1884, 15, p.322
32 Review of 1889, *Electrician*, 3 January 1890, p.224
33 *Catalogue* of the Edison & Swan United Electric Light Company, dated 1 October 1890
34 Byatt, I.C.R.: *The British Electrical Industry 1875-1914*, Oxford, 1979, pp.175-6
35 Letter from J.W.Swan to John Worsnop, quoted in an article by Worsnop about Lord Armstrong, *Newcastle Daily Journal*, 27 December 1900

36 Taylor, Robert S.: 'Swan's electric light at Cragside', *National Trust Studies,* 1981, pp.26-34 and *Proceedings of the annual conference on history of electrical engineering,* IEE, 1979
37 *The Engineer,* 17 January 1881
38 Crompton, R.E.B.: *Reminiscences,* p.97, and Bowers, Brian: *R E B Crompton,* Science Museum, 1969
39 *The Times,* 16 January 1883
40 Letter: Thomson to Preece, 13 February 1882, IEE Archives SCMs 22/523
41 Reference 38 above, p.122
42 *Electr. Rev.,* 24 November 1893, p.565 and 29 December 1893, pp.693-695
    I am grateful to Her Grace the Duchess of Devonshire for permission to inspect the remaining equipment, and to Dr Patrick Strange of the University of Nottingham for drawing my attention to it.
43 Obituary of A.W.Sclater, *Proc. IEE,* 1921

Early supply schemes

9.1   The first power stations

There are several contenders for the title of the first public electricity supply scheme, and the choice depends on the definition of 'public supply'. In several towns the Local Authority engaged contractors to provide electric street lighting and the contractors were free to offer a supply to premises in the streets being lit. The street lighting schemes already mentioned in Paris and London did not include a supply to private customers. The undertaking at Godalming was probably the first to do so.

Godalming, a small town in Surrey, was 'the first town which has combined public and private lighting in one commercial undertaking', according to a writer in 1888.[1] He explained that some parts of the town were lit by arc lamps, while other streets *and private houses* were lit by incandescent lamps, and

> When the occupier of a house wishes to use incandescent lamps, branch wires are connected at suitable points to the main cables and led into the house, a switch being inserted for the purpose of turning on or cutting off the entire supply at will.

There is no mention of fuses or meters in this or any other account of the Godalming installation. Probably the domestic consumers paid a flat rate per lamp per year. A boyhood recollection of that installation written in 1950 implies that there were no fuses or cut-outs:

> Born in 1868 I was 12 then 13 in November 1881 so I can remember it . . . My father was one of the first to have the light in our Shop and dining room and generally through the house . . . The lamps were much as they are now, but slipped into two brass slides like an inverted letter U. In those days we boys often had magnets to play with and the similarity intrigued me, so one day in our showroom when no one was about I took a needle to see if the electricity

would act as a magnet and held it across the base of the two sides. The needle vanished and on my thumb and finger were deep white hollows where the needle had been. It had instantly fused. This was never done again as you can imagine.[2]

The lamps described must have been Swan's first commercial pattern, with a long glass stem that slid into a holder with two brass contact strips at the sides. That type of lamp was made only for a year or two, which suggests that the reminiscence quoted above really is of the original Godalming installation.

The technical details of the first electric lighting in Godalming are fairly well documented (though the accounts are not all consistent), but there is no clear evidence for the date of the first supply to a private customer. According to Siemens there were 'eight or ten' private customers in May 1882, with a total of 57 incandescent lamps, and Godalming was the only town having private and public lights supplied from a central generator.[3]

Until the end of September 1881 the streets of Godalming were lit by gas. The charge for the year was £238 16s, which the Council considered excessive. It is not clear how Calder & Barrett, described in the local press as 'electrical engineers and contractors, of No. 154 Westminster Bridge-road, London' came to be involved at Godalming. They do not appear in London street directories nor in the membership lists of the Institution of Electrical Engineers. However, they gave a demonstration on Monday 26 September 1881 in which part of the town was lit by a single large Siemens arc lamp and several Swan lamps. By the following Friday evening the Council had accepted Calder & Barrett's tender to light the town for a year for £195, and three arc lamps were installed.[4]

The lighting created great local interest and was reported in the national press. Figs. 9.1 and 9.2, reproduced from *The Graphic*, show the Town Hall (known locally as 'The Pepper Box' or 'The Pepperpots') and High Street at night, and the generating plant.

The power source was a waterwheel already installed at Westbrook Mill, which belonged to the leather dressers R. & J. Pullman. The head of the firm, John Pullman, played a leading part in getting the electric light adopted in Godalming. He allowed his waterwheel to be used in exchange for having light in the mill and offices.[5]

Contemporary descriptions are not consistent in detail, but the generator was a Siemens a.c. machine running at 840 r.p.m. excited by a Siemens d.c. machine running at 1200 r.p.m. The excitation current was 12 amperes and there were two load circuits. One circuit supplied 7 arc lamps in series with a current of 12 amperes and a total e.m.f. of 250 V. The other supplied about 40 incandescent lamps in parallel at 40 V with a total current of 33 A. The driving power was said to be 10 horsepower, which gives a system efficiency of 58%. Presumably the generator had two armature circuits, one supplying each load circuit, but this is not made clear. Siemens certainly made generators with double-wound armatures and four brushes.

Fig. 9.1 *Town centre of Godalming early in November 1881 when the arc lighting was working in the main streets but the filament lighting had not yet been installed in the side streets*

Fig. 9.2 *Generating plant at Pullman's mill in Godalming*
The generator and exciter can just be seen on the right. Possibly these machines were like those in Fig. 6.28

The current was monitored by a Siemens electrodynamometer and the attendant had to regulate the speed so as to keep the current constant. Apparently there was only one instrument, and there is no record of which current was monitored.

The water supply did not prove either adequate or reliable, and steam power was soon introduced to supplement it. Another serious problem was that the incandescent lamps further from the generator were dim compared with those nearer because of the voltage-drop in the longer circuits. Calder & Barrett tackled the latter problem by adjusting the generator so that the most distance lamp was correctly lit, and then putting resistances in series with each of the nearer lamps.

In January 1882 Alexander Siemens and Siemens Brothers & Company took over the lighting contract from Calder & Barrett. He moved the steam engine and generator to a site in the town centre, which helped with the problem of variation between the brightness of the Swan lamps.[6]

The supply may have been extended to Charterhouse School, which possesses a photograph of electric lighting in the school taken about that time (Fig. 9.3). The school had a lecture on 4 February 1882 by Mr Fricker, an engineer employed by Siemens, with demonstrations of Swan lamps and Siemens arc lamps, but the supply was probably from a generating plant at the school.[7]

Fig. 9.3   *Photograph of the Library of Charterhouse School, taken in 1882*
Swan lamps of his early, long-necked pattern can be seen in the lighting fittings.
(Photo: Mr B R Souter, Charterhouse)

Siemens continued to run the lighting in Godalming until April 1884. The Council would have continued, but after canvassing for extra business Alexander Siemens told the Council that he could not find enough custom to justify continuing the supply. He needed a load of 400 to 500 private lights to make the system economic; he could find only about 100, which represents probably a few dozen shops and houses.

On 1 May 1884 Godalming reverted to gas lighting, but the Gas Company had further reduced their charges and, as *The Electrician* observed 'Godalming should, therefore, have Messrs Siemens Brothers & Co. in grateful remembrance'.[8]

The Derbyshire town of Chesterfield also had electric street lighting in the autumn of 1881. As in the case of Godalming, the Council had been unable to agree the terms of a new contract with the local gas company. The streets of Chesterfield remained unlit for about a month while proposals by several contractors were considered. Calder & Barrett and the Société Générale d'Electricité were among those who submitted tenders, but the contract was awarded to Hammond & Company, who had already lit industrial premises nearby. The first installation was of 22 arc lamps and 50 Lane-Fox incandescent lamps, at a cost of £885 for the first year. No private lights seem to have been supplied at Chesterfield, and the public lighting reverted to gas in 1884 when Hammond declined to renew his contract.[9]

The public electric lighting system of Norwich was expanded to supply private cutomers at an early date. The city's Lighting Committee had for some years been considering projects for improving the lighting of the market place and main street, and decided to try electric light. In January 1881 Crompton submitted a tender for lighting the market place for three months with two arc lamps. However, before the installation was completed the Council agreed to let the market place be used for a Fisheries Exhibition, and Crompton lit the exhibition with nine of his arc lamps and sixty Swan incandescent lamps. After the exhibition Crompton extended the public lighting in the street, trying various arrangements of arc and incandescent lamps. In March 1882 it was announced that 'Messrs Crompton now contemplate extensively adding to their plant, and . . . hiring out the current for incandescent lamps to all such consumers who care to use it within the area covered by their conducting wires.'[10]

The first steam-powered generating station giving a supply to the public was the Edison station at Holborn Viaduct in London. A similar station built by Edison at Pearl Street, New York, opened in September 1882, to give the first public electricity supply in America. The Holborn Viaduct plant was running by 12 January 1882, though the formal opening did not take place until 12 April. Edison had entered into an agreement with the City Corporation, through his London agent, E.H. Johnson, to light the Viaduct and adjoining streets for a three month trial period. It was understood that he could supply private customers, and in mid-April 1882 the station was supplying 938 lamps of which 164 were 16-candlepower lamps in the streets. The generating station itself had 216 16-candlepower lamps and 16 8-candlepower. There were thirty private customers with a total of 484 16-candlepower lamps and 58 8-candlepower. Two customers had only three lamps each. By far the largest private customer was the City Temple, the first church in the world to make use of public electricity supply.

The Holborn Viaduct system operated at 110 V d.c., the 16-candlepower lamps taking 0·8 A each. The total electrical load was therefore about eighty kilowatts. The single Edison generator installed initially was intended to supply up to one thousand lights. It was driven by a Porter-Allen horizontal steam engine capable of 125 horsepower. The generator and engine together with their common baseplate weighed over twenty tonnes. A similar set able to supply 1200 lights was soon added. A third, similar set was installed later in another building, together with a 250-light set to supply the small daytime load.

The main conductors were solid, semicircular copper conductors supported in insulating material within iron tubes. These were laid along a subway and insulated cables taken from the mains for street lamps and private customers. Each circuit was protected by a fuse, and consumption was measured by electrolytic meters.

The Holborn Viaduct station was closed down in 1886.[11]

Brighton has had an electricity supply undertaking continuously since February 1882. Robert Hammond had visited Brighton in December 1881 to give an exhibition of Brush arc lighting. The exhibition attracted much interest and Hammond was asked to extend the demonstration and run a circuit along some of the main streets so that shopkeepers could consider the advantages of the new light. A circuit nearly three kilometres long with sixteen arc lamps in series was demonstrated for a week from 21 January 1882. Applications were invited from people willing to pay 12s (60p) per lamp per week for lighting their premises. The response was adequate, and the Hammond Electric Light Company began supply on 27 February 1882, from dusk until 11 pm daily. Another generator was installed in the spring of 1883, and the tariff was changed to 6s (30p) per lamp per week plus 1s 6d (7½p) per carbon used.

Also in 1883 the engineer in charge at Brighton, Arthur Wright, introduced a system for operating incandescent lamps from the arc lamp circuits, which carried a constant direct current of 10·5 amperes. Wright connected groups of ten lamps in parallel. The failure of any one lamp would, of course, lead to overrunning of the remaining nine. Wright devised several electromagnetic mechanisms which automatically brought an extra lamp into circuit when one failed. By the end of 1885 the system had 1000 lamps connected, and two years later the total had risen to 1500, in addition to 34 arc lamps, all supplied from five Brush generators through over twenty kilometres of bare overhead mains.

The current in the Brighton system was kept constant by a boy whose duty was to watch an ammeter and adjust a carbon resistance shunting the field of the generator. In 1884 the carbon resistance was replaced by a liquid resistance in which electrodes were raised or lowered in a tank of water. Wright then devised an automatic regulator in which a long solenoid in the main circuit was arranged to raise or lower the electrodes directly. When the electrodes reached the limit of travel in either direction a pair of contacts closed to ring a bell, and the attendant had to raise or lower the speed of the engines, as appropriate. The automatic system maintained the current within 1% of its nominal value.

In 1887 it was decided to give a continuous supply, twenty four hours per day, and the system was converted to high voltage a.c. distribution at 1800 V reduced by transformers to 100 V for the consumers.[12]

The largest public supply undertaking to be formed outside the scope of the Electric Lighting Act, 1882, was the Kensington Court Electric Light Company. Kensington Court was a new housing development off Kensington High Street, West London. Nearly one hundred large houses were being built and linked with subways so that services could be provided without any need to break up the street. An electricity supply undertaking which did not use overhead wires and did not need to break up a public road in order to reach its customers was not affected by the Act.

Swan and Crompton had considered supplying electricity to another part of Kensington in 1883, and the Swan United Electric Light Company had been granted a Provisional Order for doing so under the terms of the 1882 Act. They did not proceed with the scheme, however, and the order was revoked. By 1886 Crompton's work in Vienna was a proven success and Kensington Court provided scope for a comparable supply system. He obtained an introduction to the architect of the estate, and the Kensington Court Company was formed on 5 June 1886 with a nominal capital of £10 000. Crompton was a shareholder in this Company, which entered into an agreement with Crompton & Co for the construction and maintenance of a generating station. In October 1886 it was agreed that No. 14 Kensington Court should be equipped as the first electrical showhouse.

Supply started on 1 January 1887, to three consumers. By the end of the year six further consumers had been connected, but the estate was not yet finished and the Company was receiving applications from residents outside the Kensington Court estate. The Company managed to obtain a Licence from the Board of Trade to supply customers outside the estate, though only after several months of painstaking negotiations.

The initial generating plant was a 35 kW Crompton generator, giving 100 V d.c. A second set was installed during 1887. Fig. 9.4 shows the station about 1890, by which time it had three 50 kW and four 100 kW generators, all driven by Willans engines. In 1888 the company was reconstituted as the Kensington and Knightsbridge Electric Lighting Co. Ltd., and began to give a supply over a much wider area. A second generating station, operating in parallel with the first, opened in Cheval Place in 1890, and there was also a battery substation in Queen's Terrace Mews.[13]

9.2 Early d.c. systems

All supply undertakings had to solve the problem of transmitting electrical energy at high voltage from the generating station to points nearer the consumers, then reducing the voltage and stabilising the voltage at the consumers' terminals. There was a great rivalry, known as 'the battle of the systems', between the supporters of a.c. and d.c. systems. The advantages of a.c. were that transformers could change the voltage as required, and tap-changing made voltage adjustment a fairly simple matter. The chief advantage of d.c. was that batteries could ensure continuity of supply when the generators were not running. It was also easier to operate d.c. generators in parallel than a.c. ones. The battle was eventually won by a.c. though only after the invention of the induction motor, which made a.c. supplies more attractive to industrial customers.

The undertakings so far described, except that at Godalming, used direct current, and it is convenient to trace the further development of d.c. systems before considering those that used a.c.[14]

Fig. 9.4 *Model of Crompton's Kensington Court generating station, showing five sets of Willans' steam engines coupled to Crompton generators*

An early undertaking using high voltage d.c. distribution to battery substations was established at Colchester in 1884 by the South Eastern Brush Electric Light Co. Ltd. The initial installation consisted of a semi-portable steam engine driving two Brush arc lighting generators, each giving about 10 A at 1800 V. The generators operated in parallel and supplied six installations of storage batteries which were charged in series. One battery installation was at the power station; the other five were in the cellars of shops in the town. Each installation had two batteries each consisting of eight groups of cells with 80 V across the terminals of each group. At any time one battery was being charged while the other was in service supplying the local load. The change-over was done automatically by a rocking switch with prongs dipping into sixteen iron cups of mercury for each battery. The battery being charged had its eight groups of cells connected in series. The battery in service had its eight groups in parallel. Both the charging mains and the supply cables to the consumers were lead-covered cables laid in brick ducts beneath the pavements. Consumers were charged either 0·5d (0·2p) per lamp per hour used or alternatively

2d per lamp per day in summer and 3d per lamp per day in winter. No public lighting was connected to the system. The promoters had hoped to get a load of 2000 lamps, but only a few hundred were ever connected. In 1886 the undertaking was closed and the equipment sold for scrap. Electricity did not return to Colchester until 1898.

Within a few months of the closure of the Colchester supply a similar undertaking was begun in London. The Cadogan Electric Lighting Company was registered in March 1887 and commenced supply in July 1888 to parts of Chelsea, Kensington, Brompton and Knightsbridge. Each consumer had a storage battery of 8 to 64 cells divided into four groups. Each cell had a capacity of 700 ampere-hours, and at any one time three groups of cells were available for use and connected in parallel to the consumer's load. The fourth group of cells at every installation was connected to the Company's charging main which linked all the consumers in a series circuit seven miles long. Each consumer's equipment had a motor-driven switch which disconnected one group of cells from the mains and connected another every few minutes. The generating station had three steam engine and generator sets giving 70 A at 500 V, and two or three generators were run in series as required. The Company never had many customers and in 1893 was taken over by the more successful Chelsea Electricity Supply Co. Ltd.

The Chelsea Company had been registered in 1884 but only commenced supply in 1889, after gaining the financial support of several manufacturing firms, including Callender's Cable Co., the Electric Power Storage Co., and the Electric Construction Co. These firms supplied most of the equipment used. The Chelsea Company used a system of battery substations which supplied consumers in a small area and were charged in series. They did not, however, give every consumer his own battery installation as the Cadogan Company had done. Customers were supplied at 100 volts and the maximum charge was £3 10s for up to 84 kWh per quarter and about 4p per kWh thereafter. The generating station had three engine and generator sets, each giving up to 75 A at 500 V. The battery substations were arranged more efficiently than in the systems described above, and at times of heavy load all the batteries could discharge into the supply mains at once.

The capacity of the undertaking was soon extended by installing d.c. to d.c. rotary convertors at the battery substations through which the charging mains could be connected directly to the local supply mains, and so assist the batteries at times of peak load. The convertors looked like generators, but had two armature windings and brush gear at each end.

The battery substations of the Chelsea Company were intended to operate unattended. The switching of the batteries was controlled by a device which exploited the fact that lead-acid batteries give off gas when fully charged. The gas from one cell was collected in a 'gasometer' which rose and operated the mechanism. However, it was only with rotary convertors that automatic and unattended operation proved reliable.

The Oxford Electric Company was registered in August 1891 and commenced supply in June 1892. The generating station supplied direct current at 1000 volts to convertor substations containing rotary convertors. Local distribution was at 110 V. The substations did contain batteries, but they were only a stand-by and for use when the load was so light that running the generators or the convertors was not economic. The system of generating high voltage d.c. and using rotary convertors in substations became known as the 'Oxford System', and was adopted by several other undertakings.

9.3   Early a.c. systems

The great pioneer of a.c. supply systems was Sebastian Ziani de Ferranti (1864-1930).[15] Ferranti was born in Liverpool, and developed an interest in science and engineering. In the summer of 1881, at the age of 17, he joined the Experimental Department of Siemens Brothers & Co. at Charlton. He was soon supervising electrical installations in different parts of the country and meeting leading engineers. In July 1882 he obtained his first two patents and his first Company, Ferranti, Thompson and Ince Ltd was formed in September of the same year. Alfred Thompson was an engineer and Francis Ince a lawyer interested in electricity. The company manufactured the Ferranti—Thomson Dynamo, which was a machine designed by Ferranti but so similar to a machine devised and patented by Sir William Thomson (later Lord Kelvin) that the Company paid royalties to Thomson. The Ferranti-Thomson Dynamo was a disc generator (see Chapter 6) whose winding was a simple zigzag rather than a number of distinct coils. At the end of 1883 the firm of Ferranti, Thompson and Ince was dissolved, and Ferranti set up in business on his own in Hatton Garden, London, manufacturing generators, arc lamps, meters, and other devices.

In the meantime the Grosvenor Gallery Company was running into technical difficulties. The directors sought Ferranti's advice, and were so impressed by him that on 13 January 1886 they appointed him their Chief Engineer.

The Grosvenor Gallery Company began when Sir Coutts Lindsay installed plant for lighting the Grosvenor Gallery in New Bond Street, London. The first installation had two semi-portable steam engines each driving Siemens generators giving alternating current at 2000 V. Lighting was provided by arc lamps in series, and the line current was maintained automatically at 10 A. Soon neighbouring residents and shopkeepers sought a supply of electricity, and these requests were met by transformers on the Gaulard and Gibbs series system. In this the primaries of all the transformers (Fig. 9.5), which were known as 'secondary generators', were connected in series. The individual loads were fed from the secondaries.

Lucien Gaulard and J.D. Gibbs of Paris first demonstrated their a.c. transmission system at an Electrical Exhibition in the Westminster Aquarium, London in 1883.[16] A contemporary account explains the reasons behind their system:

It has a two-fold object, first it aims at rendering it practicable for undertakers to supply current at the most economical potential permitted . . . ; and, secondly, it is intended to make the user independent of the producer ... If the secondary coils be made in several parts, each with independent terminals, these parts may be combined either in parallel, or compound parallel arc, or in series, according to the conditions under which the second or locally generated current, is to be employed.[17]

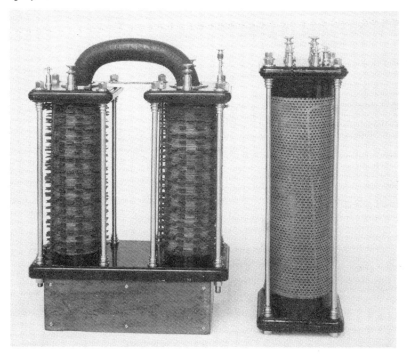

Fig. 9.5 *Two Gaulard and Gibbs transformers*
The one on the left has a closed magnetic core. The windings are flat copper rings with tabs at each end which project from the coil and are rivetted to tabs from adjacent turns. The one on the right has an 'open' magnetic circuit.

In April 1883 Gaulard and Gibbs offered to instal a demonstration of their system, free of charge, on several stations of the Metropolitan Railway Company in London. By November 1883 five quite widely spaced stations on the Circle Line (Notting Hill Gate, Edgware Road, Euston Square, King's Cross, and Aldgate) were lit electrically. The primaries of the 'secondary generators' at each station were connected in series through a main carrying a constant current of 10 A. Secondary windings supplied a mixed load of arc lamps and filament lamps at the stations. Each secondary generator had sixteen straight iron cores, arranged in groups of four, so that there were in effect sixteen transformers at each station and eighty transformers in the whole system.

The switching arrangements at each station provided that the secondaries of each group of four transformers could be connected in series or parallel as required.

The voltage at each load was controlled by moving adjustable brass sleeves which covered part of the iron core of each transformer. The system therefore had the advantage over the earlier systems that the voltage at each load was not affected by distance from the generator.

The Gaulard and Gibbs system seems to have worked well on the railway, but for reasons which are not clear the installation was not made permanent and it was removed in 1884. A system which had been seen to supply eighty secondary circuits must have seemed ideal for Sir Coutts Lindsay as he began to expand. However, each additional 'secondary generator' installed in the constant current circuit required an increase in the charging voltage, and the system soon reached its practical limit.

Ferranti rearranged the Grosvenor Gallery system for parallel working, with distribution at 2400 V and a transformer on each consumer's premises to reduce that to 100 V. He also replaced the Siemens generators by machines of his own design rated to supply 10 000 lamps of 10-candlepower, or about 350 kW. The supply frequency was $83\frac{1}{3}$ cycles per second (which is 10 000 reversals per minute).

The undertaking grew rapidly (Fig. 9.6), and it was soon apparent that the two Ferranti generators would not be adequate for long. The supply was not metered: consumers paid £1 per year per 10-candlepower lamp connected. The peak load recorded at this time was 19 500 10-candlepower lamps or equivalent.

A new Company, the London Electric Supply Corporation Ltd., was formed in August 1887 to take over the Grosvenor Gallery station and supply electricity on a far grander scale. The Corporation had an authorised capital of £1 000 000 with which to construct a power station outside London and bring electricity into the city in accordance with Ferranti's pioneering ideas. The largest shareholder was Lord Wantage. His investment of £220 000 has been described as an act of courage comparable with that which had earlier earned him the Victoria Cross in the Crimean War. Sir Coutts Lindsay invested almost £50 000.

Ferranti's plan was to build a massive power station at Deptford, on the south bank of the River Thames and then outside the built-up area of London. The site offered cheap land, easy access for coal barges, and ample cooling water. He designed generators of 10 000 horsepower (7460 kW) and planned a station to hold twelve. The most controversial aspect of Ferranti's proposals, however, was his plan to transmit power at the unprecedented high pressure of 10 000 V. He satisfied himself as to the practicability of his ideas by installing a special transformer at Grosvenor Gallery to step the 2400 volt main up to 10 000 V. There were no instruments capable of measuring 10 000 V, so he connected a hundred 100 V lamps in series across the 10 000 V terminals, and judged the brightness of one lamp to determine the voltage. The Board of Trade doubted whether the system could operate safely at 10 000 V. Ferranti said that it was quite safe with his design of main cable, which had concentric conductors with the outer earthed (Fig. 9.7). He arranged a demonstration in which a chisel was driven through a main whose inner conductor was live at 10 000 V. H.W. Kolle volunteered to hold the chisel

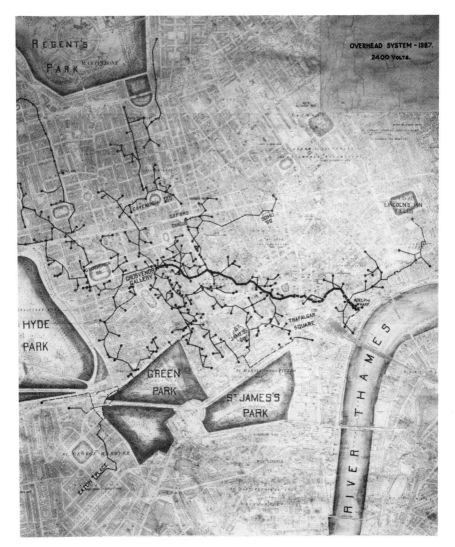

Fig. 9.6 *The Grosvenor Gallery Company's overhead distribution system in 1887, super-imposed on a modern map of London*

while another of Ferranti's assistants, C. Henty, struck it with a hammer. No harm resulted, and the Board of Trade allowed the system to be used.

Ferranti's main consisted of two concentric copper tubes of equal cross-sectional area, separated by paper impregnated with ozokerite wax. The area of copper in each conductor was 160 mm², and the main was rated to carry 250 A.

The main was laid alongside the railway lines from Deptford into London, since Ferranti thought (probably correctly) that it would be easier to get the consent of

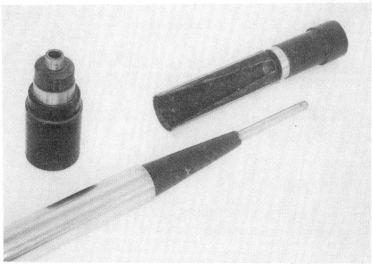

Fig. 9.7   *Ferranti's tubular main*
The main consisted of concentric copper tubes separated by wound paper insulation.
It was manufactured in short lengths with conical joints that were easily assembled
on site.

the Railway Companies than it would be to get permission from all the various
Local Authorities to lay cables under the streets.

The Ferranti main proved quite satisfactory in use and continued in service until
1933, when it was replaced by a larger cable. Short lengths of the old main were
mounted in holders as paperweights and sold in aid of the Electrical Industries
Benevolent Association.

Ferranti designed almost everything for the Deptford station himself (Figs. 9.8
and 9.9). The main building was 64 m long, 60 m wide and 30 m high. A quay
60 m long was constructed on the river. Work began on the site in April 1888 and
the Company had hoped that the station would be ready to supply current by the
end of the year. In the event it only began to supply current in 1890 and did not
take over the whole supply of the Corporation's system until February 1891.

The old Grosvenor Gallery Station could only just meet the increased demand,
and two disasters in late 1890 led to the supply being discontinued until the
Deptford plant was ready. On 15 November 1890 a man switching new transformers
into service at the Grosvenor Gallery Station drew an arc at a plug switch and
panicked: he should have pushed the plug back or opened the main circuit breakers.
The arc quickly started a fire and destroyed much of the equipment. By 26
November, however, a set of new and repaired transformers had been put in and the
supply resumed, but on 3 December a repaired transformer burnt out. Its load
transferred to others which also failed one by one. The Company, which had 312
customers at that time, decided that they had no option but to abandon the supply
until Deptford was ready.

Fig. 9.8 *Engines being assembled at Ferranti's Deptford generating station in about 1889*

Fig. 9.9 *Reconstruction of the 10 000 V to 2000 V substation at Trafalgar Square, the London end of Ferranti's main from Deptford*

In the meantime Major Marindin had been considering for the Board of Trade the special problems of electricity supply in London. His report is considered more fully below (Chapter 10), but the point which disappointed Ferranti and the London Electric Supply Corporation was Marindin's recommendation that the Corporation should only supply a much smaller area than they had hoped. In the circumstances the Directors of the Corporation decided to build only a small portion of the plant Ferranti had designed for Deptford, and not his largest planned machines. In August 1891 Ferranti resigned from the Corporation.

The Deptford Station continued to be plagued with difficulties, which probably could not have been foreseen. A particular source of trouble was the enormous boilers: 61 tubes failed and had to be replaced in 1895, for example. The quality of coal received and problems with coal handling equipment, and also with the purity of the water supply all added to the difficulties.[18]

Ferranti's Deptford scheme was an ambitious undertaking, and the subsequent development of the supply industry was along the lines he foresaw and pioneered. In moving from Grosvenor Gallery to Deptford he was stepping in the right direction, but perhaps he should have taken a smaller first step.

The public supply system is now almost entirely a.c., but for transmission over long distances d.c. systems have advantages. The high power mercury arc valve, developed mainly in Sweden in the 1930s, enables very high power currents to be rectified, transmitted over a d.c. line, and then inverted back to a.c. A d.c. cable under the English Channel has joined the British and French systems since 1962.

9.4   References

1 Pole, W., *Life of Sir W Siemens*, John Murray, 1888, p. 298
2 Letter: George S. Tanner to Mr Mealor, Borough Librarian, 12 July 1954, in Godalming Museum
3 Siemens, C.W., Evidence to House of Commons Select Committee on the Electric Lighting Bill, 27 May 1882; Questions 2429 and 2433
4 *Surrey Advertiser*, 1 October 1881, p. 5, 3 October 1881, p. 3 and *Guildford Journal*, 4 October 1881, p.3
5 Obituary: John Pullman, *The Times*, 5 April 1934, p. 12
6 Most of this section is based on the account in *Engineering*, 13 January 1882, pp. 35-6 Some details come from *The Times*, 16 December 1881, and the *Guildford Journal*, 20 December 1881, p. 3
7 *The Carthusian* (Charterhouse school magazine) 1882
8 *Electrician*, 5 April 1884, p. 485 and 26 April 1884, p. 555
9 Strange, P., 'Early electricity supply in Britain: Chesterfield and Godalming', *Proc. IEE*, 1979, **126**, pp. 863-868
10 *Engineering*, 31 March 1882, pp. 293-5
11 For the main facts of the Holborn Viaduct installation see, Parsons, R.H., *The early days of the power station industry*, Cambridge UP, 1939, pp. 10-11
    For technical details see *Electrician*, 11 February 1882, pp. 202-3, and 22 April 1882, pp. 368-9

12 Parsons, R.H., *op. cit.,* pp.12-20
    Electrician, 1887, **20**, pp. 100-102, 127-8, 152-3, 178-9
13 Parsons, R.H., *op. cit.,* pp.89-93, and Crompton, R.E.B., *Reminiscences,* Constable, 1928
14 For fuller details of the undertakins described here see Parsons, R.H., *op. cit.;* Brooke, R.C.R., 'Electricity supply in the City of Westminster', *Proc. IEE,* 1979, **126**, pp. 875-883, and contemporary descriptions in *Electrician*
15 For a full account of Ferranti's life and work, see Ferranti, Gertrude Z. de, and Ince, R., *The life and letters of Zebastian Ziani de Ferranti,* Williams and Norgate, 1934
    For a brief account see Ridding, Arthur, *S.Z. de Ferranti - Pioneer of electric power,* Science Museum, 1964
16 For a study of the Gaulard and Gibbs systems see Strange, P., 'Some tests on a Gaulard and Gibbs transformer', *Papers presented at the Weekend Meeting on the History of Electrical Engineering,* IEE, 1975, pp. 4/1 - 4/7
17 *Engineering,* 1883, **35**, p. 205
18 Smart, John E., *The Depford letter-books, an insight on S Z de Ferranti's Deptford Power Station,* Science Museum, 1976

Chapter 10

Legislation about electricity supply

10.1   The need for legislation: Playfair

Once the feasibility of a public electricity supply system had been demonstrated, a series of Official Inquiries and Acts of Parliament followed. The most important were the Inquiry conducted by Sir Lyon Playfair in 1879, the Acts of 1882 and 1888, and the Inquiry led by Major Marindin in 1889.

The promotors of electricity undertakings were making heavy capital investments, and understandably sought legal protection of their right to supply electricity in their area. The Public Health Act of 1875 had given local authorities absolute control of the layer of subsoil under the streets needed for water supply, drainage or public lighting; deeper subsoil belonged to the owners of the adjoining property. No one had any general right to dig up the streets to supply gas or electricity to private consumers. The first gas plants had been authorised by individual private Acts of Parliament, but many of the general clauses repeated in all such Acts were brought together in the Gasworks Clauses Acts of 1847 and 1871. Tramways were similarly regulated by the Tramways Act, 1870, which laid down the general conditions under which tramway undertakings could be authorised. However, a separate special Act of Parliament was still required for each undertaking, and the consent of the Local Authority was essential. An important clause in the Tramways Act gave the Local Authority power to purchase the undertaking compulsorily after 21 years. The compulsory purchase provision was apparently included to prevent private companies building up monopolies, as had happened with gas and water undertakings, and a similar provision was to be included in the early Electric Lighting Acts.[1]

In 1878 and early 1879 many Private Bills were presented to Parliament by prospective undertakers seeking powers for the supply of electricity in various parts of the country. It seems surprising, in retrospect, that so many groups were interested in opening a supply business at that time, when practical experience of arc lighting was minimal and the filament lamp was still only a dream, as yet unrealised in practice. It may be that Edison's premature announcement in October

1878, 'I have just solved the problem of the subdivision of the electric light indefinitely', stimulated proposals that had no sound technical basis.

A Select Committee of both Houses of Parliament was set up in March 1879, 'to consider whether it is desirable to authorise Municipal Corporations or other local authorities to adopt any schemes for lighting by electricity; and to consider how far and under what conditions, if at all, gas or other Public Companies should be authorised to supply light by electricity'. The chairman, Lyon Playfair, FRS, MP, later Lord Playfair of St Andrew's, (1818-98) was an eminent scientist who first trained as a doctor, then turned to chemistry and became a Professor of Chemistry before entering Parliament.

The Playfair Committee took evidence from many distinguished scientists and engineers, including William Siemens, John Hopkinson and Sir William Thomson (later Lord Kelvin). The transcript ran to over 300 pages though their report, in June 1879, was only two pages long.[2] They did not think there was a need for general legislation at the time. Electricity had advantages for lighting and 'Scientific witnesses think that, in the future, it might be used to transmit power': continuing experiment should be encouraged.

Part of the report read thus:

Unquestionably the electric light has not made that progress which would enable it in its present condition to enter into general competition with gas for the ordinary purposes of domestic supply. In large establishments the motors necessary to produce the electric light may be readily provided, but, so far as we have received evidence, no system of central origin and distribution suitable to houses of moderate size has hitherto been established.

In considering how far the legislature should intervene in the present condition of electric lighting, your Committee would observe, generally, that in a system which is developing with remarkable rapidity it would be lamentable if there were any legislative restrictions calculated to interfere with that development. Your Committee, however, are not in a position to make recommendations for conditions which may hereafter arise, but at present do not exist, as to the distribution of electric currents for lighting private houses from a central source of power. No legislative powers are required to enable large establishments, such as theatres, halls, or workshops, to generate electricity for their own use.

If corporations and other local authorities have not power under existing statutes to take up streets and lay wires for street lighting or other public uses of the electric light, your Committee think that ample power should be given them for this purpose. There seems to be some conflict of evidence as to whether the existing powers are sufficient or not. But even in regard to local authorities it would be necessary to impose restrictions upon placing the wires too near the telegraph wires used by the Post Office, as the transmitting power of the latter would be injuriously affected by the too close proximity of the powerful electric currents needed for producing light.

Gas companies, in the opinion of your Committee, have no special claims to be considered as the future distributors of electric light. They possess no monopoly of lighting public streets or private houses beyond that which is given to them by their power of laying pipes in streets. Electric light committed to their care might have a slow development. Besides, though gas companies are likely to benefit by the supply of gas to gas-engines which are well suited as machines for producing electric light, the general processes of gas manufacture and supply are quite unlike those needed for the production of electricity as a motor or illuminant.

Your Committee, however, do not consider that the time has yet arrived to give general powers to private electric companies to break up the streets, unless by consent of the local authorities. It is, however, desirable that local authorities should have power to give facilities to companies or private individuals to conduct experiments. When the progress of invention brings a demand for facilities to transmit electricity as a source of power and light from a common centre for manufacturing and domestic purposes, then, no doubt, the public must receive compensating advantages for a monopoly of the use of the streets. As the time for this has not arrived, your Committee do not enter into this subject further in detail than to say that in such a case it might be expedient to give to the municipal authority a preference during a limited period to control the distribution and use of the electric light, and, failing their acceptance of such preference, that any monopoly given to the private company should be restricted to the short period required to remunerate them for the undertaking, with a reversionary right in the municipal authority to purchase the plant and machinery on easy terms. But at the present time your Committee do not consider that any further specific recommendation is necessary than that the local authorities should have full powers to use the electric light for purposes of public illumination; and that the legislature should show its willingness, when the demand arises, to give all reasonable powers for the full development of electricity as a source of power and light.

Looking at the Playfair report with the benefit of hindsight, it is astonishing that Joseph Swan was not called as a witness nor was his work mentioned. Edison's work was mentioned briefly, but the witnesses doubted whether it would be successful. Among the experts giving evidence only John Hopkinson and Sir William Thomson seem to have foreseen the coming widespread use of the incandescent filament lamp. W.H.Preece, the 'Electrician to the Post Office', was questioned about interference on telephone lines caused by a.c. electric cables. Preece had conducted tests and found that interference was negligible if there was a separation of at least six feet (2m). Asked to look ahead Preece said he saw no reason to think that currents in electric cables would increase, nor did he think the telephone would be widely adopted by the public in Britain. It was being widely used in America, but things were different in Britain, said Preece, because 'we have a superabundance of messenger boys'.

So the Playfair Committee considered only arc lighting, and for that there was no immediate need for general legislation.

10.2   The Electric Lighting Acts, 1882 and 1888

After the Playfair Committee had reported the situation changed rapidly, largely due to the successful development and commercial production of filament lamps by both Edison and Swan. Three Acts of Parliament promoted by Local Authorities and passed in 1879 gave powers to supply electricity. The three places were Blackpool, Liverpool and Over Darwen. In 1880 similar Acts were passed in respect of Hull, Lancaster, Oldham and Irvine. Although several of these places soon had electric lighting in some streets and other public places, none gave a supply to private customers at this time. All the plans were quite modest in scale. Five of the authorities envisaged borrowing up to £5000 for electric lighting, Hull wanted £10 000 and Irvine gave no figure.

No such Acts were passed in 1881, but by early 1882 a total of twenty eight Bills before Parliament contained provisions relating to electric lighting.[3]

Seven of the bills were promoted by Companies, and they were intended to give the Companies general powers to produce and supply electricity wherever they might make an agreement with a Local Authority. The Companies would have power to break up the streets and erect poles as necessary. There were minor differences between the Bills, but it seems to have been assumed that Local Authority approval would always be required. The seven Companies were

Anglo-American Brush Electric Light Corporation
The Electric Light and Powers Generator Company Ltd.
The British Electric Light Company Ltd.
The Dublin Electric Light Company
The Electric Lighting and Synchronizing Company
Siemens Brothers & Company
Edison's Electric Lighting

Another group of Bills were eight promoted by the existing Gas Companies at Bromsgrove, Exmouth, Lincoln, Milford Haven, Northwich, Oxford, Rugby and Ventnor. The general purpose of these Bills was to permit the Companies to supply electricity as well as gas, and on similar terms. The remaining thirteen Bills were by Local Authorities who sought powers to supply electricity.[4]

The general view of the Board of Trade (or at least of the author of a Board Memorandum on the subject) was that the seven Company Bills were basically reasonable in their provisions but some of the other Bills either lacked adequate provisions for regulating the supply of electricity or created unacceptable monopolies.[3] A general Act was needed to lay down provisions which should apply to all schemes for the supply of electricity. The Board of Trade specified ten general principles.

(i)   Every person should be free to make and use electric power on his own premises without interference, so long as he does not injure or annoy others.

(ii)   Every person should be free to supply electric power to others without interference, so long as he can do it without interfering with the streets or with the property of others, or with existing electric systems, or otherwise injuring other persons.

(iii) Every person to whom electric power is supplied from any public source should be free within his own premises to use it in whatever manner he thinks best.

(iv)   In each town the local authority should have the option of establishing a general public source and distribution of electric power.

(v)   If they do not establish it themselves, then they should be empowered to license private undertakers to do it.

(vi)   In the case of concessions to private persons, the period of the concession should be limited to a short period, with a power to the local authority to resume.

(vii) Such provisions concerning force and supply to be introduced as present knowledge enables Parliament to frame.

(viii) Provisions should be introduced for protecting the Post Office and other public or private interests from injury.

(ix)   If the local authority decline to supply themselves or license others to supply, no power to use the streets, &c., should be granted without the authority of Parliament, and the powers so granted should be subject to all the conditions to which licensees of the local authority are subject.

(x)   Concessions to existing gas companies must be watched and guarded with great jealousy, because it will be their interest either to extend their present monopoly and powers of charging to the new source of light, to which they have no claim, or to defeat the practical success of the electric light.

The President of the Board of Trade, Joseph Chamberlain, introduced the Bill which was to become the Electric Lighting Act, 1882, on 3 April 1882. Chamberlain (1836-1914) was born in London but moved to Birmingham where he made a fortune in screw manufacture. He became active in local politics and then entered Parliament as a Liberal in 1876. He was a firm believer in the merits of Municipal enterprise, as opposed to either private or state monopolies, and this belief is seen in the favoured position of the Local Authorities under the Electric Lighting Act. His elder son Austen became Chancellor of the Exchequer in 1903, and his younger son Neville was Prime Minister from 1937 to 1940.

Described in its own preamble as 'An Act to regulate and facilitate the supply of Electricity', the 1882 Act envisaged three alternative procedures for establishing electricity supply undertakings: Licence, Provisional Order, and Special Act.[5] The Board of Trade could grant licences to any Local Authority, company or person, and such licences would be for periods not exceeding seven years but could be renewed. The consent of the Local Authority had to be obtained first. When the consent was not forthcoming (and it hardly ever was) the Board could grant a 'Provisional Order', which had to be confirmed by Parliament. The Provisional Order would be for up to 21 years, but any undertaking established under a Provisional Order (or under a Special Act incorporating the detailed provisions of the main Act) could be purchased compulsorily by the Local Authority after that time. The purchase price would be 'fair market value' at the time of purchase 'without any addition in respect of compulsory purchase or of good will or of any profits which may or might have been or be made . . . '. Translated this means that the purchase price would be the value of the plant as a kit of parts, not the value of the business as a going concern.

The Act gave effect to the principles laid down in the Board of Trade Memorandum. It gave a statutory right to people in the area covered by a licence or provisional order to receive a supply, and it forbade 'undue preference' between consumers. The erection of overhead mains was prohibited unless the Local Authority gave consent. A supply undertaking could avoid the provisions of the Act, and in particular the Local Authority's right to purchase, if it could either obtain consent for the erection of overhead mains or run its mains along private property such as the railways. The best known and most important example was Crompton's undertaking at Kensington Court, described in the previous chapter, where the mains were laid in subways that linked all the houses on the estate. A few companies operated in these ways, but the majority of prospective undertakers did not proceed, possibly because financiers would not accept compulsory purchase after only 21 years under the terms laid down in the Act.

Initially, after the passing of the Act, there had been a rush for licences and provisional orders, but few of the proposed schemes actually materialised. In 1883 there were 106 applications for Provisional Orders, 23 from Local Authorities and 83 from Companies. As a result 69 orders were granted and 47 applications were either withdrawn or refused. (The figures do not add up because some of the applications included districts of more than one Local Authority and had to be divided; 17 orders were eventually granted in respect of 7 applications). Ten licences were also granted, three to Local Authorities and seven to Companies. In 1884 there were only four applications in all, and only three in each of the years 1885 and 1886. The boom which had seemed imminent in 1882 had vanished, and in no case were any of the powers granted under the various Orders and Licences actually being exercised.[6]

The Local Authority purchase right was considered the main factor discouraging investment, but it may not have been the only one. There was general industrial recession in Britain in the mid 1880s, and in view of the rapid developments being made in electrical technology, there was probably also a reluctance to invest in expensive capital equipment that could soon be obsolete.

In the case of the tramways the Local Authority purchase right was a real disincentive to electrification when that became possible around 1890, but the circumstances were different. Many horsedrawn tramways were then approaching the end of their 21 year period under the Tramways Act of 1870. Financiers would not invest money which they had little hope of recovering in the time remaining. The London County Council began to buy up the various London Tramways in 1896, and ran its first electric trams in 1903, 15 years after the first trams ran in the USA.[7]

In November 1884 an influential deputation, Sir Frederick Bramwell, R.E.B.Crompton, Robert Hammond, J.S.Forbes, and other leading figures in the electrical world, called on Joseph Chamberlain. They urged him to repeal the purchase clause, and to make some other changes to the legislation. A committee was set up to draft amendments, and the result was a Bill introduced in 1886 by the distinguished scientist Lord Rayleigh (1842-1919). The intention of this Bill was to abolish the purchase clause and give electricity undertakings the same security of tenure as gas companies. Viscount Bury introduced another Bill which increased the purchase period to 24 years and provided that any undertaking purchased compulsorily should be valued as a going concern. The Government then introduced a third Bill, and after prolonged debate this was passed into Law as the Electric Lighting Act, 1888.

The new Act was brief, having only five clauses. Its effect was to retain the Local Authority's right of purchase but to extend to 42 years the period after which the purchase right could be exercised. Furthermore, the purchase price was to be 'fair market value at the time of purchase, due regard being had to the nature and then condition of such buildings, works, materials, and plant, and to the state of repair thereof, and to the circumstances that they are in such a position as to be ready for immediate working'. In other words the undertaking was to be valued as a going concern, not just as a collection of buildings and machinery. The new Act also provided that the consent of the Local Authority was necessary before the Board of Trade could grant a Licence, although the Board was empowered to dispense with that consent if the Local Authority refused unreasonably. It also clarified a point which was not clear under the 1882 Act: a Licence or an Order did not confer any monopoly. Another provision brought existing overhead lines under the control of the Board of Trade.

The 1888 Act was followed by a rapid growth in electricity supply schemes. By January 1889 there were 26 central generating stations operating in Britain and 17 others under construction.[8] By January 1890 the number was 46[9] and by January 1891 it was 54, of which 17 were in London.[10]

10.3 Electricity in London

London was different from the rest of the country. It had the largest concentration of population, and probably the wealthiest potential customers. It is not suprising, therefore, that while many towns had one or two supply undertakings London attracted more than a dozen. Until the formation of the London County Council, which came into being on 1 April 1889, the metropolis had a large number of autonomous, small Local Authorities with no body responsible for the London area as a whole.

Early in 1889 the Board of Trade announced that no more Licences would be granted for the London area. Prospective undertakers would have to seek an Order. The Board then set up a public inquiry to consider the applications and related issues. The inquiry opened on 3 April 1889 and was conducted by Major Marindin, assisted by Major Cardew. The inquiry closed on 1 May and its report was published on 20 May. Eight Companies and thirteen Local Authorities were represented, and many leading electrical engineers gave evidence.

The report was well received by all concerned, who seem to have accepted that Marindin had analysed the situation accurately and produced fair and reasonable proposals. The report laid down the lines on which electricity supply in London could develop, and its main conclusions were:

(i)   everybody should be able to have an electricity supply
(ii) all the Orders for London should, as far as possible, have the same general conditions
(iii) Companies and Local Authorities should work under the same statutory obligations
(iv) there was no objection to a Company having powers over a large area provided it had sufficient capital to meet the demand
(v)   competition between supply undertakings was desirable, though in cases where the Local Authority objected to any company only one should be allowed. Where two undertakings supplied the same area, they should not both use alternating current (because at that time there were no satisfactory, a.c. motors available).
(vi) all overhead lines should be removed and cables laid underground.

The report specified the areas to be allotted to the seven companies then operating in London: the London Electric Supply Corporation, the Metropolitan Electric Supply Co., the House-to-House Electric Light Supply Co., the Notting Hill Electric Lighting Co., the Kensington and Knightsbridge Electric Lighting Co., the Chelsea Electricity Supply Co., and the Westminster Electric Supply Corporation.

10.4   Power companies

The first supply undertakings could operate only within the boundaries of one local authority. As the industry developed the advantages of supplying larger areas soon became apparent. The Acts of 1882 and 1888 had not envisaged this development, and consequently further Private Bills had to be promoted in Parliament.

Four Private Bills concerned with electricity supply were promoted in 1897. The Chelsea Electricity Supply Co. wanted powers to purchase land compulsorily for a power station. The Metropolitan Electric Supply Co. wanted to lay mains between their area of supply and a new power station right outside their area. The Central Electric Supply Co., a new Company with no supply area of its own, wanted to purchase a power station site and lay mains to take a supply to authorised undertakers. The General Power Distribution Co., a syndicate of large manufacturers around Chesterfield, sought approval for a scheme to supply electricity over a large area of the Midlands to authorised undertakers.

The Chelsea and Metropolitan Bills were approved quickly, but the others raised quite new points of principle. A joint Committee of both Houses of Parliament was set up in 1898 under the Chairmanship of Lord Cross to consider the issues. It reported that 'where sufficient public advantage is shown' powers should be granted to supply electricity over an area that included several local authorities, from generating stations 'of exceptional dimensions and high voltage'. They also stated that any purchase clause would be quite inappropriate where several local authorities were involved.[11]

Despite the views of the Cross Committee no general legislation providing for power companies was enacted at that time. Several power companies obtained private Acts of Parliament. The Durham, North Metropolitan, Lancashire and South Wales Companies obtained powers in 1900, and within a few years about twenty power companies had been established covering most of the rural parts of the country.

As a result of municipal opposition most towns were excluded from the areas of the power companies, although it would probably have been in the financial interests of consumers to join in. The progress of the electricity supply industry before the First World War was bedevilled by political arguments about the merits of municipal enterprise. Chamberlain had been a keen advocate, and so long as the optimum size of a supply system was smaller than a local authority area there was no problem. The new power companies, however, were a threat to the power of local authorities, and through the powerful Association of Municipal Corporations pressure was brought on all Members of Parliament whatever the part of the country concerned. With the return of a Liberal Government in December 1905 and a strong campaign in the press in 1906, the climate of opinion swung in favour of the companies. The new President of the Board of Trade, David Lloyd George, introduced the Bill which became the Electric Lighting Act, 1909, and gave effect to the recommendations of the Cross Committee by permitting the Board of Trade to authorise the establishment of power companies without further recourse to Parliament.

London continued to be a special case. It was the only area in which the Board of Trade had permitted undertakings to compete. (There was no legal monopoly in other areas, but in practice the Board only granted one order or licence in a town). The County of London Electric Power Bill was introduced in 1905 with the object of unifying the capital's generating arrangements. The proposal was that three new generating stations should be erected on the banks of the Thames to supply the existing authorised distributors and a few large users over an area a little larger than the then County of London. The Bill was eventually defeated, but the evidence produced in its support was impressive. There were twelve local authorities and fourteen companies supplying electricity in the area. The capital invested was £3 127 000 by the local authorities, £12 530 000 by the companies. Their total generating capacity was 165 000 kW and their annual output 160 000 000 kWh, which gives a load factor of only 11%. The average cost of production was 1·55d (0·65p) per kWh and the average charge to the customer 3·16d (1·32p) per kWh. Yet some of the existing undertakings were supplying large power consumers at below 1·0d per kWh, implying that there was considerable scope for reducing charges. There was also a large volume of potential business untouched. The total motive power installed in London factories was 600 000 horsepower, but only 40 000 were electric.[12]

The engineer behind the County of London Bill was Charles Merz (1874-1940), who was also largely responsible for the pioneering scheme based on Newcastle upon Tyne, which was already bringing an efficient and economical electricity supply to the industrial north east of England. In the course of discussion about the London scheme, Lloyd George remarked to Merz that electricity supply in London was not a matter of engineering, but of politics.[13] London had to wait until after the First World War for a coherent electricity supply system.

10.5 A national system of electricity supply

After the First World War the British Government decided that some more central-ised organisation had to be imposed on the electricity supply industry. During the war the installed generating capacity had increased by 39% (from 1120 MW to 1555 MW) but the number of units sold had increased by 106% (from 1318 m per year to 2716 m.). The biggest new load was electric arc steelmaking. The lesson was clear: wartime controls had increased the efficiency of the supply industry. Furthermore, electricity supply was becoming increasingly significant in British industrial and commercial life. Several bodies reported to the Government on the need for post war reorganisation, and the consensus was that 'a new and inde-pendent Board of Commissioners free from political control and untrammelled by past traditions' should be established with extensive powers. The Electricity Supply Act, 1919, established the Electricity Commissioners with the duty of 'promoting, regulating and supervising the supply of electricity'. The Act was a step in the right direction, but the Commissioners' powers proved inadequate.

They established Joint Electricity Authorities whose role was to encourage co-operation between the separate undertakers in an area, but the Electricity Commissioners could only act by persuasion, their powers were minimal. One such authority was the London and Home Counties Joint Electricity Authority, set up in 1925. The Authority did not achieve its aim of unifying charges throughout its area before being taken over at nationalisation in 1948. It did, however, succeed in halving the average cost of electricity, and it reduced the wide variations between the charges of the undertakings in its area.[14]

During the 1920s there was pressure on the British Government to establish a transmission network linking the whole country, so that electricity could be generated in the most efficient stations and conveyed wherever it was wanted. The cost of generation at that time varied enormously from place to place and between different power stations. It was clearly in the consumers' interest to be linked to the most economical stations. An additional advantage was that interconnection of stations increased the security of supply, since if one generator or one station suffered a breakdown its load could be supplied from elsewhere.

In early 1925 the Government set up a committee under the chairmanship of Lord Weir to 'review the national problems of the supply of electrical energy' and to report on 'the policy which should be adopted to ensure its most efficient and effective development'. The committee sought technical reports from several independent experts and reported in May 1925. They proposed the establishment as quickly as possible of an independent body, to be known as the 'Central Electricity Board' with the duty of constructing a 'gridiron' system of high-voltage transmission lines. The gridiron, which soon became known as the National Grid, was to interconnect certain selected power stations and the existing distribution systems.

The Weir Committee recommended that the operation of the selected stations should be left to the existing authorised undertakings but that the new Board should have overall control. The Board was to purchase the electricity generated in the selected stations and resell it to authorised suppliers. The non-selected stations would rapidly be closed down. The Committee thought that 58 stations should be selected (43 existing and 15 new ones) and that 432 existing stations should be closed. The Government accepted the report and its recommendations were incorporated in the Electricity (Supply) Act, 1926.

The Central Electricity Board outlined a preliminary plan for the grid in their first Report, which dealt with the Board's activities up to the end of 1928. It was a scheme for interconnecting areas, not a scheme for point-to-point transmission of power. They adopted 132 000 V as the standard for primary transmission lines, with secondary transmission at 66 000 and 33 000 V. Before construction could begin, they had to design conductors, insulators, pylons, protective systems, and control arrangements.

An eight-year programme for constructing the grid was announced in 1927. By the end of 1935, when the whole of Britain was linked except for north-east England, there were 4600 km of primary transmission lines in operation, and 1900 km of secondary transmission. By 1946 these had been increased to 5900 km and 2400 km respectively.

The biggest problem facing the Central Electricity Board was the standardisation of frequency. In their first Report they noted that three-phase 50 Hz a.c. supplies were standard throughout Europe, except in Italy. In Britain 77% of the installed capacity of authorised undertakers in 1926 was 50 Hz plant, mostly three-phase, and the Board adopted that standard. The main exception was the north-east England area around Newcastle upon Tyne, which was supplied at 40 Hz. Glasgow, Birmingham, South Wales and London had some 25 Hz supplies. It was realised that the cost of changing the frequency of 23% of the national supply system would be considerable; when completed in 1947 it proved to be £17·5m.

A further task which faced the Board was the standardisation of supply voltage. A committee under the chairmanship of Sir Harry McGowan investigated this question in 1935-6. They found that of 642 supply undertakings 282 supplied a.c. only; 77 supplied d.c. only; and 283 supplied both a.c. and d.c. There were 43 different supply voltages ranging from 100 to 480 V. The real problem was that, after agreeing on a standard supply voltage, much customers' apparatus would have to be changed or converted. It was generally agreed, however, that a uniform supply voltage was highly desirable, and until it was achieved domestic appliance manufacturers had to make and stock a range of models designed for different voltages. In the case of radio and television sets with an internal transformer, a tapped primary was provided on the transformer so that the set could be adjusted to the local supply voltage. It was only in 1945 that a standard (240 V a.c.) was finally laid down. After the Second World War the British Government decided to take the whole electricity supply industry into public ownership. The Bill was introduced into Parliament in December 1946 and became effective on 1 April 1948. Under the new organisation the British Electricity Authority was established to exercise a general co-ordinating function and control the policy and financial structure of the industry as a whole, except for the north of Scotland. The Authority was responsible for the generation of electricity and its bulk transmission to fourteen separate statutory Area Electricity Boards.

In taking control of the power stations and the grid system, the new Authority also became responsible for running the seven grid control districts, each with its own grid control room. In order to give the best possible service, it was decided to divide the management and operation of the power stations and the grid into 14 Generating Divisions, corresponding as closely as possible with the Area Electricity Boards.

The Electricity Act, 1957, further changed the structure of the industry by abolishing the British Electricity Authority and establishing the Central Electricity Generating Board, responsible for the power stations and grid, and the Electricity Council. The chief functions of the Electricity Council were 'to advise the Minister on all matters affecting the industry and to promote and assist the maintenance and development by the Generating Board and the Area Boards of an efficient, co-ordinated and economical system of electricity supply'. The Council was also given special responsibility for finance, research, and industrial relations.

The growth of the electricity supply industry is illustrated in Table 10.1.

In most other countries the electricity supply system has remained in private ownership, though with a considerable degree of government control. In the United States, for example, there were still over 3800 plants supplying electricity for public use in 1947. On the other hand, the federally created Tennessee Valley

Authority attained a capacity of over 2 400 000 kW in 1948.

Table 10.1   *Electricity Supply in England and Wales*

Year*	Net generating capacity MW	Number of customers Millions	Units sold per year Millions
1895	79		38
1900	295		120
1905	775		450
1910	960		1000
1915	1300		1700
1920	2400	0·9	3240
1925	3900	1·7	5040
1930	6200	3·5	8160
1935	7300	7·0	13200
1940	8700	9·6	21900
1945	11100	10·0	28000
1950	11500	12·0	38000
1955	17300	14·1	58000
1960	25500	15·5	86000
1965	34400	17·0	130000
1970	46900	18·3	168000
1975	58500	19·3	196000
1980	57000	20·3	205000

*Calendar years to 1945, then the 12 months ending 31 March of the year specified. Source: Electricity Council Publications. The figures should be treated with caution since the basis of calculation has varied from time to time, but they show the general pattern of progress.

10.6   References

1   Farrer, T.H., Secretary to the Board of Trade, letter to *The Times* 11 September 1884
2   House of Commons Sessional Papers, 1879, **11**, p.375, 'Report from the Select Committee of Lighting by Electricity'
3   Memorandum on Electric Lighting Bills of 1882, dated 12 February 1882, PRO BT/13/13
4   The thirteen Local Authorities were Aberdeen, Blackburn, Bolton, Dundee, Glasgow, Greenock, Hull, Macclesfield, and Manchester, who already supplied gas, and Accrington, Chedderton, Newcastle upon Tyne, and Bodiham and Hopton, who did not supply gas
5   For a contemporary commentary on the 1882 Act see Poley, Arthur P., and Dethridge, Frank: *A handbook on the Electric Lighting Act, 1882 . . .*, Simpkin, Marshall & Co, and Waterlow & Sons Ltd, 1882
6   The figures in this paragraph are derived from the annual report of the Board of Trade to Parliament in respect of proceedings under the Electric Lighting Act. House of Commons Sessional Papers: 1883, **64**, p.165; 1884, **72**, p.197; 1885, **71**, p.165; and 1886, **60**, p.227
7   *London County Council Tramways Handbook,* The Tramway & Light Railway Society, Worthing, 1970

8 Review of 1888 in *Electrician*, 4 January 1889, pp.250-253

9 Review of 1889 in *Electrician*, 3 January 1890, pp.218-221

10 Review of 1890 in *Electrician*, 2 January 1891, pp.264-266

11 'Report of Joint Select Committee on Electrical Energy', *House of Commons Sessional Papers*, 1898, 9, p.615

12 Garcke, Emil: 'Electricity supply – the London problem', *The Times*, 30 May 1906

13 Quoted in Rowland, John: *Progress in power – the contribution of Charles Merz*, Merz & McLellan, Newcastle upon Tyne, 1960

14 *The story of the London and Home Counties Joint Electricity Authority 1925-1948*, published by the Authority, 1948

Chapter 11

Nationwide electricity supply

11.1 Power stations

The electricity supply industry whose beginnings were discussed in Chapter 9 has prospered, despite the legislative problems described in Chapter 10. By 1900 it was apparent that the advantages of a.c. systems outweighed those of d.c. for most purposes, though d.c. supplies continued in some areas for another sixty years.

The most progressive electricity supply undertaking in Britain at the turn of the century was Newcastle upon Tyne Electric Supply Company (NESCO) which had its own Act of Parliament as a power company and gradually took over undertakings covering a large area of industrial north-east England. One of the undertakings NESCO acquired was the Newcastle and District Electric Light Co., which had begun supply from its Forth Banks Power Station in 1890. The station was equipped with two 75 kW turbogenerators, designed by Parsons and supplied by Clarke, Chapman, Parsons, & Co. This station had the distinction of being the first to use steam turbines rather than reciprocating steam engines.

The practical steam turbine is due to Sir Charles Parsons, who constructed his first turbogenerator in 1884. That machine (Fig. 11.1) ran at 18 000 r.p.m. and generated 75 A at 100 V. The Hon. Charles Algernon Parsons (1854-1931) was a younger son of the third Earl of Rosse, himself a distinguished scientist and a President of the Royal Society. After two years at Trinity College, Dublin, Parsons went to Cambridge in 1873. He graduated in 1877, was apprenticed to Sir William Armstrong & Co., and then joined Clarke, Chapman & Co., of Gateshead, where he took charge of the newly established electrical department. He used his mathematical ability to solve two design problems: the mechanical problem of shafts rotating at very high speeds, and the problem of distributing the drop in steam pressure and the corresponding increase in steam volume along the length of the machine. He obtained his first turbine patent in 1884, the essential principle being that steam was expanded in graded stages through successive sets of blades arranged along the axis of the turbine. The turbines supplied for the Forth Banks Station were successful, but before further orders could be obtained a difference of opinion between Parsons and his partners led to the partnership being dissolved. Parsons established a new firm, at Heaton on the outskirts of Newcastle, under the name of C.A. Parsons & Co. Ltd. He wanted to make turbines and generators, but by his original agreement the patent for Parsons' turbine belonged to his former partners. Parsons resolved the problem by designing a turbine in which the steam expanded radially through sets of blades arranged on rings of ever increasing diameter. This did not infringe his earlier patent arrangement in which the steam flowed axially through the turbine.

The first radial flow turbines were installed at Cambridge in 1892. Cambridge Corporation had obtained electric lighting powers but they decided to transfer their powers to a company set up for the purpose, the Cambridge Electric Supply Co Ltd. Parsons was managing director. The power station had three 100 kW turbo-generators running at 4800 r.p.m. The initial load was 230 10-candlepower lamps, or about 14 kW. The sets were mounted on rubber blocks, not bolted rigidly to the foundations, and the quiet, vibration-free running that resulted attracted further orders from areas where residents objected to the noise and vibration of reciprocating engines.[1]

The vibration of reciprocating engines was a major problem. C.H. Spurgeon, the Baptist preacher, lamented the nuisance at his orphanage in Stockwell caused by the generating station of the City and South London Railway.

Alas, the Electric Railway is doing us terrible damage by three engines fixed, 400 horsepower each, just against wall of girls' house. They intend putting 3 more, and already they cause the houses to vibrate like ships at sea. I fear the law will give us no real remedy. I pray about it, and God can do more than the courts.[2]

Some injunctions were in fact obtained against supply undertakings, and the advent of the turbine was warmly welcomed. Similar machines to those at Cambridge were installed at Scarborough and Portsmouth, and then in 1900 Parsons supplied two single-phase 1500 kW sets to Elberfeld, Germany. These were by far the largest generating sets installed anywhere in the world at that time, and they gave an impressive demonstration of the economies of large turbines. The Cambridge sets had required 12·7 kg of steam per kWh generated; the corresponding figure for the Elberfeld sets was only 8·3 kg.

There was a demand for turbines far beyond the manufacturing capacity of Parsons' works in Newcastle. He therefore sold his European rights to the Swiss engineering firm of Brown, Boveri & Cie and his American rights to the Westinghouse Company.

A few power stations were built using reciprocating engines even after the supremacy of the turbine had been established. The NESCO built a 2100 kW station at Neptune Bank, Newcastle, which was opened in June 1901 by Lord Kelvin. The engineer responsible was Charles Merz. He chose slow-speed reciprocating marine-type engines because he was anxious to convince the industrialists of Tyneside, many of them shipbuilders, that the supply he was offering them would be completely reliable. Before the end of the year, however, Merz was planning to enlarge the station and instal turbines.

The last power station to be built with reciprocating engines was Greenwich station, opened in May 1906 to supply the newly electrified London tramways. The station had four 3500 kW sets and the original plan was to add four more, but wiser counsels prevailed and turbines were installed. The reciprocating engines, obsolete before they were even built, were finally scrapped in 1922. Only a few reciprocating engines of this size were ever built. Most engineers preferred to have a larger number of smaller machines. It was then possible to vary the number of engines in use to suit the demand, and those engines that were running could be operated at maximum efficiency.

The choice of prime mover for a power station affects the generator design. The relatively slow reciprocating engines required a multi-pole generator to give alternating current at a reasonable frequency. This follows from the fundamental relationship

speed of rotation = number of pairs of poles × frequency

The fastest reciprocating engines could run at 500 r.p.m., so that to generate at 50 Hz the generator needed six pairs of poles. Steam turbines in Britain and Europe usually run at 3000 r.p.m. driving a generator with a single pair of poles to give a 50 Hz output. (In America the figures are 3600 r.p.m. and 60 Hz). A few sets were made running at half the speed with a four pole generator.

Early generators all had the armature windings, in which the current was generated, rotating and the field magnet fixed. Brushes and slip rings were used to take the current from the rotating armature but as machines were made larger the current-collecting arrangements proved to be the limiting factor. The solution was to 'invert' the machine, placing the field magnets on the rotor and the armature windings on the stator. The brush-gear then only had to carry the magnetising current for the field windings. The last large rotating-armature generator was a 1500 kW three-phase machine ordered late in 1901 to expand the Neptune Bank station. This was also the first three-phase turbine set installed for public supply.

The first rotating field generators had 'salient' poles built up on the rotor shaft. Charles Brown suggested that the field windings should be carried in slots milled in the surface of the rotor, which should be a single forging. Since the speed of rotation is fixed, the maximum diameter of the rotor is determined by consideration of centrifugal forces. Brown's basic design concept for large generator rotors has been used ever since. Charles Eugene Lancelot Brown (1863-1924) was the son of an English father, also Charles Brown, and a Swiss mother. He was head of the Oerlikon company's electrical department and worked with Oskar von Miller (1855-1934) and Michael von Dolivo-Dobrowolsky (1862-1919) on high voltage power transmission in 1890. In October 1891 Brown founded Brown, Boveri & Cie with his German friend Walter Boveri.[3]

Since 1903 the history of power station development has been one of improvements in details and increases in size rather than radical changes in principles. The Carville station opened in Newcastle in 1904 had two turbogenerators designed to yield 3500 kW each, though on test it was found that they could give 6000 kW. Carville was the first station in Britain to follow the practice already established in America of having a 'system control room' with a diagram on the wall showing generators, substations, and the state of all switches in the system, with a control engineer in overall charge of the whole system. Carville used 2 kg of coal for each unit (kilowatt hour) of electricity generated. The North Tees power station at Middlesbrough, planned during the First World War and opened just afterwards. had 20 000 kW machines and consumed just under 1 kg of coal per unit generated. The steam temperature at North Tees was 370°C. The 'B' station at Dunston opened in 1930 operated at 427°C and required only 0·6 kg of coal per unit. The 50 000 kW sets at Dunston were the largest and most efficient in Britain and generated directly at 33 kV, so avoiding the need for a transformer with each generator. The 105 000 kW set installed at Battersea Power Station, London, in 1933 was for some years the largest in Europe.

Since 1933, generator sizes have increased. By 31 March 1980 the Central Electricity Generating Board, covering England and Wales, had 29 generators rated at 500 MW or more (Figs. 11.2 and 11.3). Progress has been a matter of many small refinements in plant. The average consumption of coal by 1980 was 0·49 kg per kWh generated. The total quantity of coal burnt in power stations was 80 million tonnes, from which 163 000 GWh of electricity was supplied (1 GWh = 1 000 000 kWh). In addition, 25 GWh were generated in nuclear power stations.

Fig. 11.2   *The stator of a modern 660 MW generator being assembled*
(Photo: NEI Parsons Ltd.)

Fig. 11.3   *Low pressure turbine, part of a 660 MW assembly, being lowered into its housing*
(Photo: NEI Parsons Ltd.)

Oil-fired stations and a little hydropower brought the total supplied up to 222 000 GWh. Total generation figures for Great Britain at various dates were given in Table 10.1.

The first nuclear power station in Britain, Calder Hall, was opened by the Queen in 1956. That station was built and run by the Atomic Energy Authority and was more a research establishment than a commercial power station. The first two commercial nuclear power stations were in England at Berkeley in Gloucestershire, commissioned in June 1962, and Bradwell in Essex, commissioned in July 1962. By 31 March 1980 seven more were in service in the United Kingdom and three were under construction. The three under construction each have a design capacity of 1320 MW.

All nuclear power stations derive their power from the fission of a particular isotope of uranium, but there are a variety of ways of making a nuclear reactor to control and utilise the reaction. Different approaches were adopted in Britain and the USA when civil nuclear power began, because of differences in the two countries' nuclear industries. In particular America had a large uranium enrichment plant and also a desire to build nuclear powered ships. The pressurised water reactor (p.w.r), using enriched uranium fuel and ordinary water as coolant was ideal for use on a ship because it is reasonably small and operates at a high power density. This type of reactor was therefore developed in America for power stations also. Britain had no enrichment plant, and so wanted a reactor that used natural uranium. Such a reactor needs a moderator, to slow down the neutrons released by fission, and the choice lay between heavy water and graphite. Graphite was chosen, and carbon dioxide was used as the coolant. The result was the Magnox reactor, named after the magnesium alloy used to clad the fuel elements. The Magnox reactor operates at a lower power density than the pressurised water reactor, and is therefore larger in size and dearer to build, but since it uses natural uranium the fuel cost is less.

The Magnox stations are still in service, but from the mid 1960s enriched uranium became available in Britain. The Atomic Energy Authority then designed the advanced gas-cooled reactor (a.g.r,) and built a small prototype at Windscale. Subsequently several a.g.r. stations were ordered by the Central Electricity Generating Board. At the time of writing the CEGB has announced its desire to build a p.w.r. station at Sizewell, where there is already a Magnox station.

Nuclear power generates strong feelings, and the CEGB's proposals for building further nuclear power stations are the subject of much public debate. On the one hand, it is argued that nuclear power is a proven technology offering electricity cheaper than any other fuel. On the other hand, it is said that nuclear power involves unacceptable risks, and the problems of disposing of radioactive waste have not all been solved.

There are other possible sources of power besides the fossil fuels and nuclear energy. There is little scope for hydropower in Britain, although it is important in other countries. In the European Economic Community the total generating capacity of 243 000 MW (1977 figures) is provided by 183 000 MW fossil fuel, 33 000 MW hydro, 19 000 MW nuclear, and 8000 MW pumped storage.

Pumped storage systems require two lakes at different levels. When the demand for electricity is low, then water is pumped to the higher lake; when demand is high, then the water is allowed to flow back driving a turbine to generate electricity. Usually the motor/pump and generator/turbine are the same machine. Pumped storage systems are popular with electricity supply authorities because electricity itself cannot be stored in significant quantities, and pumped storage helps to even out the demand on the generators. Furthermore, a pumped storage system can be started very much more quickly in response to load fluctuations than can a steam turbine-generator. To run an electricity supply system efficiently the number of generators actually turning at any moment has to be matched to the demand. At the Grid Control Centres careful attention is given to such things as weather forecasts and the timing of popular radio and television programmes which all affect people's habits. If a million people all switch a kettle on at the end of a programme, it adds up to an enormous jump in demand. Generating capacity that can be brought in quickly is therefore very valuable. The CEGB have a 360 MW pumped storage plant at Ffestiniog, commissioned in 1961, and are building a 1500 MW plant at Dinorwic, due to be in service in 1983.

Tidal power, making use of the rise and fall of the tides, is an attractive sounding idea that is very difficult to realize in practice. A barrage across the Estuary of the River Severn was first suggested about 1900, though the necessary civil engineering works would be so expensive it is unlikely ever to be worthwhile. There is one tidal power scheme in existence, on the River Rance in France. Because of the general geography of the area the difference between high and low tides at the mouth of the Rance is greater than almost anywhere else in the world. A barrage has been built across the river which contains 24 units capable of operating either as motors and pumps or as turbines and generators. The total capacity is 240 MW.

Wave power, extracting the energy of the waves, seems a more promising prospect than tidal power. Various devices have been designed to be driven by the action of the waves and then to turn a generator. Wind power may prove worthwhile for isolated communities, and at the time of writing a 1 MW wind generator is proposed for the Orkneys. Research into all these possibilities is continuing, on a modest scale.

11.2   Conductors

When electricity supply began those manufacturers who had been supplying telegraph cables took up the manufacture of power cables. They knew that copper was the best metal to use for the conductor and that it should be stranded to give increased flexibility. Vulcanised rubber and gutta percha were the established insulating materials for telegraph cables, but they proved less satisfactory for heavily loaded power cables because they softened when the cable became warm, and the core tended to 'flow' downwards under its own weight. Many insulating materials were tried out for power cables at the end of the nineteenth century. The most successful was vulcanised bitumen, generally known as v.b.[4]

W.O. Callender, a manufacturer of road-surfacing material, had the idea that the bitumen he imported from Trinidad might provide the basis of a material for insulating cables. His son William, a chemist, carried out a series of trials that led to v.b. The name is a little misleading since less than half the material is actually bitumen. V.B. was patented, and in 1882 William and his eldest brother Thomas Octavius Callender (1855-1938) established the Callender Bitumen Telegraph and Waterproof Company at Erith, Kent. V.B. cable quickly became very popular in many countries, and Callenders established a factory in America as well. V.B. cables proved reliable, at least for low voltages.

In most countries, underground cables were much preferred to overhead wires, both on aesthetic grounds and for safety, though overhead wires were often used in country areas to save money. British and European engineers usually preferred to bury their cables directly in the ground, while in the USA the use of conduits was preferred. Callenders developed an intermediate system in which v.b. cables were laid in an iron trough and the trough was then filled with bitumen (Fig. 11.4). 'Callender's solid system', as it was called, gave the cable excellent mechanical protection, and there were many variations, such as using wooden or earthenware troughs rather than iron to save cost.

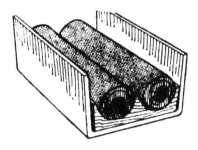

Fig. 11.4 *Callender's solid system*
Cables with vulcanised bitumen (v.b.) insulation and laid in troughs of iron, wood or earthenware which were then filled with hot bitumen

Another important insulating material for cables was jute impregnated with oil or resin. Such cables operated satisfactorily at higher voltages than were possible with v.b., but the upper limit was only about 2500 V. Above that voltage the only possible insulation was vulcanised indiarubber, until S.Z. de Ferranti introduced paper insulation.

The earliest cables were either single core, with two cables laid together, or two core, with the cores running side by side in a single sheath. With the development of a.c. supplies and the increase of telephone services induction in telephone wires became a problem. An American engineer, R.S. Waring of Pittsburgh, developed 'anti-induction' cable, with concentric conductors. From 1887 such cables were made in Britain by the Fowler-Waring Company.

Although most early electric power distribution was by cable there were important exceptions where bare conductors were used, supported on suitable insulators. The most important was a system developed by Crompton in which copper strips were stretched across glass or porcelain insulators in a duct, usually under the pavement (Fig. 11.5). The system was first used at Kensington Court, and subsequently in several British towns, including London and Brighton. It was simple and reliable. Some Crompton strip main was in service in Northampton until 1979.[5] It could be uprated by laying additional copper strips on top. Service cables to individual customers were clamped on to the strips. Its disadvantage was that the insulation resistance was low because the insulators became dirty, and the ducts were liable to become flooded. Similar systems were installed by other British engineers and were also used extensively in Paris.

When S.Z. de Ferranti became engineer of the Grosvenor Gallery Company in London in 1886, he took over a distribution system using rubber insulated cables working at 2400 V. When he conceived his scheme for a large power station at Deptford transmitting at 10 000 V into central London, he experimented with both rubber and jute insulation. Neither was satisfactory. The rubber had too high a permittivity, so that the cable took an excessive charging current; the jute tended to catch fire. He designed a main of his own, the 'Ferranti tubular main', described in Chapter 9.

Fig. 11.5   *Reconstruction of Crompton's copper strip main*

Paper insulation had been used before. The Norwich Wire Company had been founded in the USA in 1884 to make both telegraph and electric lighting cables insulated by a helically wound paper tape. J.B. Atherton visited America in 1889, saw the Norwich Wire Company's products, and obtained the British rights. On his return to England he formed the British Insulated Wire Company, later merged with Callenders in BICC, to manufacture paper-insulated cables (Fig. 11.6). Ferranti was invited to join the board of the new company.

Fig. 11.6 *10 kV paper insulated, lead sheathed, stranded, concentric cable made by the British Insulated Wire Company in 1896*

Ferranti designed a flexible cable similar to his tubular main. The inner conductor was a stranded copper wire and the outer conductor was flat strips of copper arranged in a circle. Fifteen miles of this cable were supplied to the London Electric Supply Corporation in 1896.

For cables with parallel (rather than concentric) cores Ferranti pointed out the advantages of non-circular cores with a uniform thickness of insulation. Cables such as his 'clover-leaf' cable require less insulating material and are therefore smaller and cheaper than they would be if the cores were circular.

With three-phase cables it was recognised from the start that if each core was provided with adequate insulation for the voltage between that core and earth, then the insulation between cores would be excessive. A construction was adopted in which the three cores carried only enough insulation to withstand the voltage between cores. When the three cores had been laid up together a 'belt' of insulation was put round the whole assembly to bring the core-to-earth insulation up to an adequate thickness (Fig. 11.7). Such 'belted' cable became standard practice for voltages up to 33 kV.

After the First World War the power companies demanded higher voltage cables, and manufacturers produced belted cables rated at 66 kV and designed to have the same electric stresses as the successful 33 kV ones. It soon became clear that something was wrong with the new cables, which suffered a number of breakdowns, but the cause was not at first apparent. Analysis of the electric field pattern in a three-core cable showed that the belt was carrying a component of electric stress parallel

Fig. 11.7   *'Belted' three-core cable*
The insulation around each core is adequate for half the core-to-core voltage. A further belt of insulation around all three cores increases the thickness to that necessary for the core-to-earth voltage

to the layers of paper, and since paper is electrically stronger through its thickness than along its surface that was thought to be the cause. Further research showed, however, that this was only a secondary factor. The primary cause was that in large, heavily loaded cables the cores expanded when they got warm. As they cooled and shrank spaces would be left in the insulation and, at the voltages being used, discharges would occur in the spaces and burn the insulation. Trouble arose when the maximum stress in the insulation reached about 4 kV/mm.

The solution to the problem of the 66 kV cables was to provide an earthed screen of metal or metallised paper around each core, so that the entire electric stress was borne by a homogeneous insulation. The economic advantage of belted construction had to be given up. From 1928 onwards 'screened' or 'H-type' (after the inventor Martin Hochstadter) construction was used for all cables of 33 kV and above.

For even higher voltages and powers it was necessary to avoid the creation of voids in the insulation as the core expanded and contracted. In 1930, the Callender Company introduced a cable with an oval conductor. The idea was that as the cable got hot the core would change into a more nearly circular form but the cable sheath would not be stretched; on cooling the cable sheath should return to its original form, without voids forming. The real solution, however, was the oil-filled cable, originally developed in 1920 by Luigi Emanueli, chief engineer of the Italian Pirelli Company. He used a thin mineral oil which flowed through a duct formed by a steel helix in each core. The oil permeated the whole structure of the cable so that any void which formed was immediately filled with oil. Oil reservoirs above the level of the cable maintained the oil pressure and took up any expansion or contraction.

The dielectric strength of oil-filled cable increases with increasing pressure. C.E. Bennett made use of that relationship for his 'oilostatic cable' introduced in 1931. By that date techniques for making long lengths of welded steel pipeline were well established. The oilostatic cable was an essentially conventional paper-insulated screened cable, without an outer sheath, drawn into a steel pipeline which was then filled with oil under pressure. The first experimental installation operated at 66 kV at Philadelphia, and in 1935 a 135 kV oilostatic cable carrying 100 MVA was installed for the Pennsylvania Railroad at Baltimore.

Some oil-filled cables were made 'self-compensating'. The sheath, which had to be specially designed for any pressurised cable, was given some degree of flexibility and the cable acted as its own reservoir, without the need for external reservoir stations.

Gas-filled cables were the subject of much experiment. There were several options: the paper insulation could be dry and no impregnating oil used or the paper could be impregnated in the conventional way and the gas used to maintain pressure. In the latter case the gas could either be in direct contact with the oil or separated from it by a flexible diaphragm. During the 1930s several gas-filled cables were laid in Britain and a few in other countries, but the lengths involved were small.

Cables with plastic insulation, usually p.v.c. based, were introduced for low and medium voltage service in the 1950s, and are now used almost universally. The impregnating oils and waxes used in very high voltage cables have been the subject of much research, the trend being to replace resin and resin oil by mineral oil products. The choice of copper for the conductor was long almost unquestioned, though for a brief period, about 1906 to 1913, aluminium was cheaper than copper and some aluminium cables were made. The choice is not only governed by the cost of the metal. The electrical conductivity of aluminium is only 60% that of copper: consequently, aluminium cables are larger than the equivalent copper cables and require more insulating material. Because aluminium always has a thin but tough oxide coating it is difficult to make good electrical connections with aluminium cables. However, since 1950 the price of copper has risen much more than the price of aluminium and cable makers and the electricity supply industry have undertaken extensive research on the problem of jointing aluminium cables. Aluminium cable is now widely used in local distribution networks.

In 1926 work began on the task of linking the supply systems of the whole country through the national grid. This required 4600 km of 132 000 V circuit, which was possible only by using overhead transmission lines. The Board adopted steel-cored aluminium conductors 20 mm in diameter carried on steel pylons; each circuit could carry 50 MVA. Aluminium was used, rather than copper, because aluminium conductors are only about half the weight of copper conductors of the same current-carrying capacity.

In 1949 the British Electricity Authority decided that the capacity of the grid would have to be increased. They began planning a 275 kV system, known as the Supergrid, with a capacity of 500 MVA per circuit, which was expected to meet all requirements for at least twenty years. The first section, 66 km of single-circuit line, was brought into service in 1953. In the USA the Hoover Dam Project, in the 1930s, had utilised 287 kV on a 480 km transmission line, so the technology for handling such high voltages was well established before it was needed in Britain, where the distances are shorter. In 1980 the Central Electricity Generating Board had just over 5000 km of 400 kV transmission line and over 1700 km of 275 kV line. In addition there were 65 km of 400 kV cable and 520 km of 275 kV cable.[6]

11.3   Switchgear

Breaking a circuit carrying a current is a far more complicated matter than may appear at first sight. Even a domestic light switch will usually produce a small spark as the contacts open, and for this reason 'snap-action' switches were developed early in the history of electricity supply (Chapter 13). The problems are far more serious in switches controlling very large voltages and currents. The spark is created because as the contacts begin to move apart the areas of contact decreases, the contact resistance rises, and the last areas of contact become heated and then melt. Finally a tiny portion of the metal evaporates, forming a plasma, and then the surrounding air ionizes also. In larger sizes, a switch becomes known as a circuit breaker and the 'spark' becomes an 'arc', but the difference is one of size and nomenclature rather than physical principle.

The arc which forms when a circuit breaker opens is an aid to the breaking of the circuit, not, as might be supposed, a hindrance. If the current were 'chopped' immediately the contacts opened, then enormous surge voltages would be induced because of the self-inductance of the circuit. The arc performs the vital function of allowing the current to continue to flow until the instantaneous value of the current falls to zero. This occurs 100 times per second in a 50 Hz system, or 120 times per second in a 60 Hz system. At the point when the current becomes zero, the arc extinguishes itself naturally; for satisfactory operation, the medium in which the arc occurs must have cooled sufficiently to prevent the voltage rise in the following half-cycle from restriking the arc. A circuit breaker for d.c. is more difficult to make than an a.c. breaker because it has to depend on the arc being drawn out to a length which the prevailing voltage and current conditions cannot sustain.

The early power stations were controlled by simple knife-switches on slate or marble panels. Even the simplest such switches had main and auxiliary contacts. The main contacts carried the current (or most of it) when the switch was closed.

The auxiliary contacts were linked to the main ones by springs and they were arranged to open with a snap action after the main contacts, even if the switch were operated slowly. In this way the arc drawn between the contacts as they opened was broken as quickly as possible, and burning and pitting of the contact surfaces was confined to the auxiliary contacts, which could easily be replaced.

As voltages increased the switchgear was placed behind the panel so that it was impossible for the operator to touch any live parts. Subsequently remote control was adopted, which is totally safe for the operator and also makes it possible for the controls for switches in several different locations to be grouped in a central control room.

As currents increased switching became more difficult. The first refinement, introduced before 1900, was the 'magnetic blowout'. This depends on the fact that an electric current (including an arc) experiences a sideways force as it passes through a magnetic field. In a circuit breaker with magnetic blowout the conductor carrying the current is arranged so that the arc drawn when the contacts open is forced sideways. Consequently the arc takes a longer path through the air than it would otherwise do, and it is extinguished more quickly.

During the first decade of the twentieth century, oil-immersed circuit breakers were introduced, in which the contacts are separated in a tank of oil (Fig. 11.8). The function of the oil was first to cool the arc, thus assisting its extinction, and then to increase the insulation between the contacts so as to prevent the voltage which appears after the circuit is broken from restriking the arc.

Two main types of circuit breaker were developed during the 1920s and 1930s: the air-blast and oil-blast breakers. In the air-blast breaker a fast-moving jet of air under high pressure is blown against the arc so as to stretch it out into a thin column which loses heat rapidly. The magnetic blowout circuit breaker can be used for circuits up to about 16 000 V; a single pair of air-blast contacts can interrupt a 100 000 V circuit. For higher voltages several pairs of contacts have to be connected in series and operated by a linked mechanism.

In oil-blast breakers the contacts are enclosed in oil as in the simple oil-immersed breaker. The heat of the arc evaporates some of the oil, which then dissociates into carbon and a large volume of hydrogen at high pressure. The pressure of the hydrogen is used to draw out the arc, usually against a series of baffles. Hydrogen has a very high thermal conductivity, which makes it a good coolant for the arc. The main disadvantage of oil-filled circuit breakers, especially the oil-blast type, is the risk of fire if a defective breaker should fail under pressure and explode.

During the 1940s the heavy, incombustible gas sulphur hexafluoride was introduced as an arc-extinguishing medium for circuit breakers. The general construction of the breakers was similar to the air-blast breakers, but the physical properties of sulphur hexafluoride make it more suitable than air for the purpose. However, although sulphur hexafluoride itself is inert, both sulphur and fluorine are highly corrosive substances — fluoride especially so. This restricts the materials that can be used in the construction of a sulphur hexafluoride breaker. The gas has to be kept perfectly dry, and so the breaker cannot easily be opened for inspection.

In addition, the gas liquefies at about 10°C under the operating pressures used, and if the breaker is for outdoor use it must be heated. The main applications for sulphur hexafluoride circuit breakers have been indoors in situations where the reduction of fire risk is important.

Fig. 11.8   *11 kV oil-immersed circuit-breaker used by the Clyde Valley Electric Power Co., in 1904*

Vacuum switches in which the contacts separate in a vacuum have been used for low currents since about 1930. Switches rated at up to 15 A were available in 1934, and vacuum reed switches rated at 0·5 A or less have been used extensively in telephone exchanges. Since 1965 vacuum circuit breakers have been used at ratings up to 6·6 kV and 2 MW for motor controllers. They have been found to be reliable in service and require less maintenance than other types.[7]

As the supply industry expanded in the early years of the twentieth century the importance of large switchgear became increasingly apparent. A particular problem was that there was no facility for testing circuit breakers, which made it difficult to evaluate different designs. The first short-circuit testing station was built at Hebburn-on-Tyne in 1929 by A.C. Reyrolle & Co., under the direction of H.W.

Clothier. The company had been established in London in 1897 by A.C Reyrolle (1864-1919), a French engineer, and transferred to Hebburn in 1901.[8]

The problem of circuit breaker testing was one of the matters selected for special attention by the Research Committee set up by the Institution of Electrical Engineers in 1912.[9] When the Electrical Research Association was created after the First World War it devoted part of its resources to switchgear, and much basic research into air-blast circuit breakers was undertaken at the ERA laboratory at Leatherhead in the 1930s (Fig. 11.9).

Fig. 11.9 *Experimental air-blast circuit-breaker used by the ERA in the 1930s*
The circular windows in each part of this two-pole breaker enable the action to be filmed with a high speed camera

11.4 Transformers

Faraday's ring with which he demonstrated electromagnetic induction in 1831 (Chapter 2) is sometimes called the first transformer. Since the essence of a transformer is two coils of wire linked by a 'magnetic circuit' Faraday's device is rightly called a transformer, but Faraday himself had no concept of a.c. circuits linked by the device he had made.

Pixii's first generator in 1832 (Chapter 6) produced alternating current, but that was regarded as an aberration until Ampère showed how to rectify the output. No one suggested connecting the output of Pixii's machine to Faraday's ring. Joseph Henry in America conducted experiments similar to some of Faraday's and discovered 'self-induction', the phenomenon that can cause a spark at the moment a simple battery circuit is broken. Henry also showed with his experimental coils that he could transform an *intensity* current into a *quantity* current, or transform a *quantity* current into an *intensity* current, by varying the numbers of turns in his coils.[10] The expressions *intensity* and *quantity* were commonly used to distinguish between electrical effects of relatively low e.m.f., and effects of relatively high current.

The induction coil makes use of both the effects described in the previous paragraph. The current in a first coil, the primary, is broken repeatedly by an interrupting device; transformer action is used to generate a high e.m.f. in another coil, the secondary. Several people made induction coils in the mid-19th century, though the best known was Rühmkorff, who was particularly active in the 1850s. Heinrich Daniel Rühmkorff (1803-1877) was a German instrument maker living in Paris. He appears not to have discovered any new principles himself, but he studied the ideas proposed by others and applied his own skill as an instrument maker to the manufacture of very good induction coils.[11]

The general principle of operation of a transformer was well understood by the time alternating currents came to be used in practice. The first people to use transformers on a large scale were Gaulard and Gibbs, who used 1:1 transformers with their primaries connected in a series circuit to the generator and loads connected to individual secondaries. As described in Chapter 9, Ferranti converted the Grosvenor Gallery system to the modern arrangement in which transformer primaries are connected in parallel at the distribution voltage. At the same time as Ferranti was advocating his system for the Grosvenor Gallery Company the Ganz Company in Hungary was working on similar lines. They demonstrated a transmission system using transformers at the Hungarian National Exhibition in Budapest in May 1885.[12]

The introduction of transformers and a.c. distribution in America was largely due to Westinghouse, who purchased two Gaulard and Gibbs transformers and acquired the patent rights in the USA. George Westinghouse (1846-1914) is best known for his air brake, and his early work was mainly connected with railways. In 1884 he turned to electric lighting, and later made generators for producing hydroelectric power from the Niagara Falls.

Ferranti's earliest transformers were wound on a bundle of strips of sheet iron whose ends were then folded round, interleaved and clamped to form the core (Fig. 11.10). By 1891 he developed a design in which the coils were preformed and sheet iron strips were then threaded through, bent round, and clamped. For higher powers several bundles of iron were used, with air spaces left between them for cooling (Fig. 11.11). Also by 1891 transformers with cores built up from stamped iron laminations were being made by the Brush company, to a design of Mordey's (Fig. 11.12).

Fig. 11.10   *Ferranti transformer of 1888*
The core is strips of iron sheet bent round, interleaved, and clamped by the wooden frame

The transformers used in the early a.c. distribution systems had, by modern standards, very high iron losses. In 1896 the distribution system in Newcastle upon Tyne lost 40% of the power generated. Sheffield lost 30% and even Bedford, which apparently had the most efficient a.c. network at the time, lost more than 10%. Most of these losses were hysteresis and eddy current losses in the transformer cores, and a particularly unfortunate feature was that the losses increased with age. The 'ageing' process could increase the core losses by 50% in two years, and some undertakings used to withdraw transformers from service periodically, dismantle them and re-anneal the cores to restore their original magnetic properties.[13]

Research during the later 1890s showed that heating and mechanical forces increased the rate of ageing, but little improvement was obtained until silicon steels were developed by Sir Robert Hadfield (1858-1940) in the early years of the twentieth century. The new alloy, known as 'Stalloy', was available in quantity by 1906, and quickly adopted for transformer cores. A surprising feature of the new alloy was that its magnetic properties actually improved with age, so that an old transformer could be more efficient than an apparently identical new one. As an apprentice and as an engineer in Sheffield Burnand made tests on Hadfield's steels from 1895 on, and made measurements on the Hadfield transformers now in the Science Museum.

The cores of modern power transformers are still made from steel alloys based on Hadfield's work, and the efficiency of a large modern transformer is well above 99%.

Fig. 11.11   *High power Ferranti transformer of 1891, rated at 10 000-2000 V*
The core is 10 bundles of iron similar to those in Fig. 11.10, and with air spaces
between them for cooling

Fig. 11.12   *Mordey transformer of 1891 made by the Brush Company*
The core is assembled from iron stampings, a few of which can be seen loose in
front of the tranformer, and the whole assembly is clamped by long bolts

The modern high capacity transformer connecting the National Grid is a substantial feat of mechanical enginering. A transformer rated at 100 MW and with an efficiency of 99·5% has 0·5 MW of losses to be dissipated as heat. Forced circulation oil-cooling is usually employed and, since the oil is also part of the insulation, it must be kept clean and dry as it is pumped through the system. The mechanical forces between the windings and the frame and between one winding and another can be very high under fault conditions, so the mechanical construction has to be extremely strong. The construction problems are further complicated by the fact that a large grid transformer is too heavy to be transported in one piece, and must therefore be assembled on site.

11.5 References

1   For a survey of early power stations and C.A. Parsons' contribution see Parsons, R.H.: *The early days of the Power Station Industry*, Cambridge UP, 1939, especially pp. 170-183
2   Letter: C.H. Spurgeon to Charles Joseph, 10 February 1891, in Spurgeon's College, London
3   Obituaries of Brown originally in German but quoted in English in Vrethem, Åke T: *Jonas Wenström and the three-phase system*, Royal Institute of Technology, Stockholm, 1980, pp. 26-28
4   For the early history of cables see Hunter, P.V., & Hazell, J.T.: *Development of power cables*, London, 1956
5   *Electr. Times*, 16 November 1979, p.3
6   Annual Reports of Electricity Council and Central Electricity Generating Board
7   Baker, M.J.: 'Vacuum Contactors', *Electr. Times*, 6 January 1978
8   Obituary of Reyrolle, *Electr. Rev.*, 1919, 84, p. 270, and *Electrician*, 1919, 82, p. 268
9   Minutes of the Committee, Archives of the IEE
10 Henry's work is summarised in MacLaren, Malcolm: *The rise of the electrical industry during the nineteenth century*, Princetown University Press, Princetown, New Jersey, 1943, pp.25-31 and 164-167
11 MacLaren, *op. cit.*, pp.165-167
12 *Electr.. Rev.*, 8 August 1885
13 *Electrician*, 22 November 1895, p. 107 and p. 124; 22 May 1896, p. 116; 2 June 1899, p. 192

Chapter 12

Electrical measurements

12.1 Electrostatic measurements

Simple measurements of electrostatic charge were attempted in the 18th century. The first published scientific paper on electrical measurements was probably a letter from an unknown writer addressed to John Ellicott, FRS, and read to the Royal Society on 6 March 1746.[1] The writer had been studying the interaction of light-weight electrified balls hanging on threads, and noted that they behaved 'with such a Constancy and Regularity, as to the Effect, that I apprehend one may thence deduce a Gauge or Standard, whereby to measure electrical Powers'. He then described a simple balance with which he measured an attractive force of 200 grains weight (13 g) produced by a charged body. Two years later Ellicott wrote a long article describing his own experiments, in which he was seeking 'the Laws of Electricity', though he did not achieve any quantitative success.[2]

In 1786 the Rev Abraham Bennet wrote to the Royal Society describing an 'Electrometer'.[3] It was in fact the gold-leaf electroscope, with two pieces of gold leaf hanging in a glass cylinder and electrically connected to a flat brass plate at the top.

Accurate electrostatic measurements may be dated from 1785 when Charles Augustin de Coulomb (1736-1806) invented the torsion balance electrometer. This had a long silk thread supporting an horizontal straw covered with sealing wax. The straw could rotate inside a large glass cylinder and the silk thread was fixed to a rotatable cap at the top. A metal ball within the glass cylinder but connected to an external knob could be charged and would attract the wax on the straw. According to Coulomb the device was so sensitive that he could measure a force of 10^{-4} grain (6·5 mg).[4] With this instrument he showed that the force between electrostatic charges varies inversely as the square of the distance between them.

The English scientist Sir William Snow Harris made a simple device for weighing the attractive force between two charged discs in 1834. Sir William Thomson improved Snow Harris' device in 1855, adding guard rings around the discs which eliminated edge effects. With Thomson's device the voltage, force, area of effective plate and spacing could be related mathematically.[5] Subsequently Thomson became interested in electrostatic instruments to measure high voltages. His 'quadrant electrometer' (Fig. 12.1) was sold commercially from 1887, and could read up to 10 000 V. It has two vertical sheets of metal with a narrow space

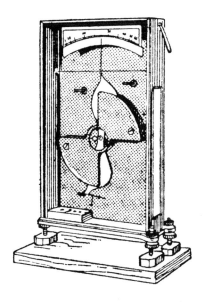

Fig. 12.1   *Quadrant electrostatic voltmeter designed by Sir William Thomson (later Lord Kelvin) in 1887, for measuring up to 10 000 V*
Electrostatic attraction between the fixed and moving vanes is balanced by gravity

between them in which a pivoted metal vane was drawn by electrostatic attraction. The motion of the vane was restrained by gravity, and it carried a pointer. In 1890 Thomson designed his 'multi-cellular' instrument (Fig. 12.2) in which a number of 'quadrant' movements were coupled together. In this instrument the moving assembly was suspended on a torsion wire, which provided the necessary restraint. The instrument resembled a carriage lamp and is sometimes called the 'carriage lamp voltmeter'. It was used for voltages of a few hundred. Both patterns of Thomson's electrostatic voltmeter are usually known as Kelvin instruments, although Thomson did not become Lord Kelvin until after he had completed the basic development. Many were made by the firm of Kelvin & James White Ltd.

The electrostatic voltmeter has been modified for use as a Wattmeter. For this the instrument requires two sets of fixed vanes, one pair occupying the first and third quadrants of a circle, the other pair occupying the second and fourth. The moving vane and one pair of fixed vanes are connected across the load whose power consumption is being measured, and the other pair of fixed vanes is connected to a resistance in series with the load, on the side of the load to which the moving vane is connected. The deflection of the instrument is then proportional both to the voltage across the load and the voltage across the resistance. The latter is a measure of the current being taken, so the instrument reads the product of the load voltage and current, or power. This arrangement was used in the cable industry in the early years of the 20th century for measuring dielectric loss in cables.

Fig. 12.2   *Thomson's (Kelvin's) 'Carriage lamp voltmeter' of 1890 has a number of quadrant assemblies on a common shaft, and the deflecting force is balanced by a torsion wire*

12.2 Measurement of current

Most electrical measurements are made by using the magnetic force of a current in a coil of wire to move a pointer across the scale. The magnetic force acts against a controlling force which may be produced by gravity, a spring, or another magnet.

Oersted's discovery of the effect of the current in a wire on a compass needle, and Schweigger's multiplier (Chapter 1) formed the basis of the galvanometer. The French physicist Claude-Servain Pouillet (1791-1868) began the precision measurement of current with his tangent galvanometer (Fig. 12.3) and sine galvanometer, both first described in 1837. In both instruments the coil is mounted in a vertical plane, the needle pivots freely, and the controlling force is provided by the earth's magnetism. In the tangent galvanometer the coil is made of large diameter relative to the length of the needle, which is pivoted at the centre of the coil. Pouillet showed that if the plane of the coil lies north-south then the tangent of the angle of the deflection of the needle is proportional to the current in the coil. In the sine galvanometer the coil initially lies north-south but is rotated about a vertical axis to bring it into line with the needle. With this arrangement the current is proportional to the sine of the angle of deflection. The sine instrument is the more accurate of the two, though the tangent galvanometer is simpler to use.

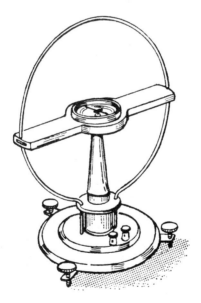

Fig. 12.3 *Pouillet's tangent galvanometer of 1837*
The instrument has to be aligned in the earth's magnetic field. The magnetised needle is deflected by the field of the current in the large loop and restrained by the earth's field. Provided that the magnet is short, the current is proportional to the tangent of the angle of deflection of the magnet

The electrodynamometer was invented in 1841 by Wilhelm Eduard Weber (1804-91), who was following up Ampère's mathematical analysis of the forces between current-carrying conductors. The principle of the electrodynamometer is that a movable coil, usually supported by a torsion suspension, hangs within a fixed coil. When current flows in both coils magnetic forces are created which tend to twist the suspended coil. Usually the suspended coil is returned to its initial position by twisting the upper end of the suspension, the necessary angle of twist being the reading of the instrument. Dynamometer instruments are suitable for current and power measurements. For current measurements both coils carry the current being measured and the angle of deflection is proportional to the square of the current. This arrangement was used for monitoring the current in the first supply at Godalming. For power measurements one coil carries the current and the other is connected across the load. The instrument is equally suitable for a.c. and d.c. measurements. Dynamometer instruments were manufactured commercially by Siemens & Halske from about 1880 (Fig. 2.4) and were used as laboratory instruments after direct-reading, but less reliable, meters had been introduced.

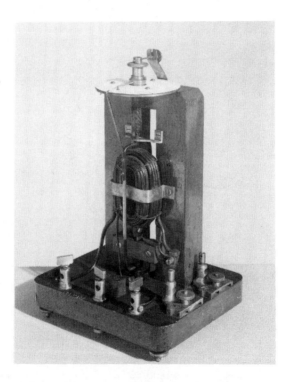

Fig. 12.4   *Siemens electrodynamometer, for measuring heavy currents*
To take a reading the coil is returned to its initial position by twisting the fixing of the torsion suspension, and the angle of twist is read from the scale at the top of the instrument

The Kelvin Ampere balance, in which the attractive force between fixed coils and coils on a balance is compensated by a sliding weight on a balance arm, is similar in principle to the electrodynamometer. The usual construction has six flat coils, two on the opposite ends of a moving beam and four fixed, one above and one below each of the moving coils. In 1894 two such balances, specially designed for the purpose and calibrated by electrolytic deposition, were designated legal standard instruments and held by the Board of Trade. The legal ampere was defined thus:

A standard of electrical current denominated one ampere being the current which is passing in and through the coils of wire forming part of the instrument marked "Board of Trade Ampere Standard Verified 1894" when, on reversing the current in the fixed coils, the changes in the forces acting upon the suspended coil in its righted position is exactly balanced by the force exerted by gravity at Westminster upon the iridio-platinum weight marked A, and forming part of the said instrument.

Sir David Salomons noted with satisfaction that the accuracy was within one fifth of one per cent.

A galvanometer needs to be calibrated by some means if it is to indicate voltage or current directly. The need for calibration can be avoided if the galvanometer is used simply to determine the absence of current. This is the principle upon which potentiometer and bridge measuring circuits are based.

The potentiometer was described by Poggendorff in 1841.[6] A constant current through a uniform wire stretched along a scale gave a potential between two points proportional to their spacing, provided that no current was taken from the points. The potentiometer was calibrated by connecting a standard cell between two points on the wire and observing the reading of a galvanometer in series with the cell. When there was no deflection of the galvanometer the voltage between the points was equal to the voltage of the standard cell. The value of an unknown voltage could then be found in terms of the standard by finding the length of potentiometer wire across which the voltage was equal to the unknown voltage.

The bridge circuit (Fig. 12.5) was described by Wheatstone in a long paper on electrical measurements presented to the Royal Society in 1843.[7] In the paper Wheatstone clearly attributed the circuit to S.H.Christie, and called it the 'Differential Resistance Measurer', but it has been known ever since as the 'Wheatstone Bridge'. Christie devised the circuit and method for a purpose similar to but not quite the same as Wheatstone's. His object was to compare the electromotive forces induced by magnetoelectric induction in different metals. He knew that when two coils, one of iron wire and one of copper wire but otherwise identical, were connected to a galvanometer different currents would flow when a magnet was plunged into or withdrawn from the coils. The question he sought to answer was whether the difference in current was due entirely to the difference in resistance of the two metals or whether a different e.m.f. was induced in each. To answer the question he connected the two coils in series so that their e.m.f.s were in opposition. He found that no current flowed and that therefore the same e.m.f. was induced in each coil. The credit is due to Christie for the important concept of comparing two e.m.f.s. by putting them (or part of them) in opposition and making some adjustments until exact equality is indicated by a zero reading on a galvanometer. Wheatstone's achievement was to convert Christie's laboratory procedure into a practical instrument, and to publicise it.

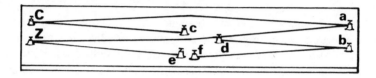

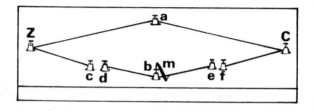

Fig. 12.5   *Wheatstone's original drawings of the bridge circuit, from his Measurements paper of 1843*

The bridge can also be regarded as a development from the potentiometer. Two resistances in series with each other are connected in parallel with the potentiometer across a suitable supply. A galvanometer connected between the junction of the resistance and a movable contact on the potentiometer wire enables the point to be found at which no current flows through the galvanometer. The junction and the movable contact are then at the same potential and the ratio of the resistances is the same as the ratio of the two portions of the potentiometer wire, which can be read from the scale. If one resistance is known, the other can be calculated.

The bridge and potentiometer are excellent laboratory instruments but not convenient for rapid, practical use. The need was for a galvanometer to indicate current and voltage directly, and which was of more manageable size than Pouillet's tangent galvanometer and did not need to be aligned with the earth's magnetic field every time it was used. A simple galvanometer, in which a magnetic needle is deflected by the effect of the current and restrained by a weight or spring, depends on the stability of the magnet. Until the development of better magnetic steels around the turn of the century, all magnets were liable to deteriorate quite rapidly, so galvanometers could not be calibrated with a permanent scale.

Wheatstone appreciated that 'it would greatly facilitate our quantitative investigations if we had a certain and ready means of ascertaining what degree of the galvanometric scale indicated half the intensity corresponding to any other given degree'. He described a switching arrangement which connected the galvanometer into a circuit either directly or through a resistance network. If the galvanometer has a resistance R, then the network consists of another resistance R in parallel with the instrument and a resistance ½R in series with the combination (Fig. 12.6). The network plus galvanometer has the same resistance between its terminals as the galvanometer alone, so the external circuit is unaffected when the network is introduced but the current through the galvanometer is reduced to one half of its previous value.

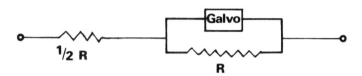

Fig. 12.6   *Wheatstone's circuit for doubling the range of a galvanometer without affecting the test circuit*

He then showed how the circuit of a galvanometer could be changed so that it carried currents of $\frac{1}{2}$, $\frac{1}{3}$, $\frac{1}{4}$, etc., of some initial value. The corresponding currents could be observed and the instrument calibrated for use as a direct reading instrument. Wheatstone also showed how to calculate the value of shunt resistance to be used with a galvanometer so that it could be used to measure currents of any desired magnitude. These circuits are the basic circuits used in the current and voltage ranges of all multi-range meters.

An important early development in galvanometers was the 'astatic needle', first described by Leopoldo Nobili (1784-1835) of Florence. In this arrangement two magnetic needles are fixed parallel to one another on the same shaft. One needle is inside the coil and one outside, and the needles point in opposite directions. The result is that the magnet field of the coil twists both needles in the same sense, but the earth's magnetic field tries to turn them in opposite senses. The effect of the earth's field is therefore reduced, and may be eliminated if the needles are of equal strength.

Wheatstone used galvanometers with astatic needles, and also fitted a microscope to enable him to read the instrument more accurately.

12.3   Direct reading instruments

The rapid progress of electric lighting around 1880 created a demand for practical, direct-reading instruments. Several workers made improved galvanometers graduated to read voltage and current directly, though calibration remained a problem.

The first practical moving iron ammeter was invented by Ayrton and Perry in 1879. An iron core was drawn into a cylindrical coil carrying the current to be measured, and the movement of the core drove a pointer. In their magnifying spring ammeter (Fig. 12.7) the restraining force was provided by a helical spring wound from a flat metal strip. The spring was fixed rigidly at one end and stretched or compressed by the movement of the core at the other. The change in length of the spring caused the free end to twist through a relatively large angle, and the pointer was fixed to the free end.

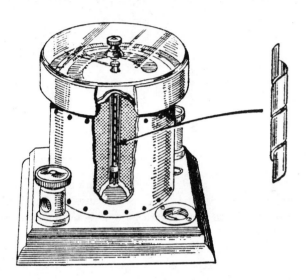

Fig. 12.7 *The Ayrton and Perry magnifying spring ammeter*
A small iron core at the bottom is attracted upwards, compressing the spring which then twists

Several other designs of moving iron instrument were developed in the 1880s. Schuckert's instrument had a movement with a strip of soft iron mounted asymmetrically on an axis in a coil carrying the current to be measured. The magnetic field caused the iron to twist.

The moving coil galvanometer, with a light coil suspended between the poles of a permanent magnet and a fixed iron cylinder inside the coil, is due to Jacques Arsène d'Arsonval (1851-1940), a medical professor of the Collège de France. When the magnet was provided with curved pole pieces, so that the air gap between the poles and the iron cylinder was uniform over a large angle, the d'Arsonval instrument gave an angular deflection directly proportional to the current.

The first large-scale commercial manufacturer of such instruments was the Weston Electrical Instrument Co. of Newark, New Jersey. Weston made the permanent magnet the frame of the instrument, wound the coil on a copper former so that its movement was damped by eddy currents, and mounted the moving system in jewelled bearings (Fig. 12.8). Two spiral springs carried the current to the coil and provided the controlling force. A further development was the 'Unipivot' movement (Fig. 12.9) devised by the London instrument maker R.W.Paul in 1903. In these instruments the moving parts are supported on a single bearing, giving lower friction and hence greater sensitivity.

The 'hot-wire' instruments make use of the fact that a wire becomes longer when heated. The first hot-wire voltmeter was designed in 1883 by Philip Cardew (1851-1910), the consulting engineer who later assisted Major Marindin with his inquiry into London's electricity supplies. In one form of the instrument three metres of platinum-silver wire were stretched over pulleys in a tube nearly a metre long. One end was fixed and the other end was held by a spring and connected by gearing to a pointer. Such instruments were often used on early power station switchboards.

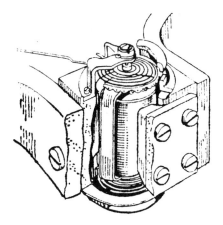

Fig. 12.8   *The Weston moving coil instrument, introduced in 1888*
The coil swings in a uniform air gap between the poles of the magnet and an iron core

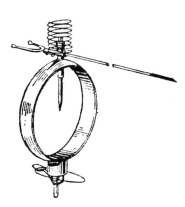

Fig. 12.9   *The Paul 'Unipivot' movement of 1903, which gives a very sensitive instrument*
The restraining force is provided by hair springs which also carry the current to the coil

In the 1890s Hartmann and Braun improved the hot wire meter by making use of the sag of a taut wire rather than its change of length (Fig. 12.10). A wire attached to the middle of the current-carrying wire was held taut by a spring and drove the pointer. The resulting instrument was far less bulky than Cardew's.

The first recording instrument was probably one made about 1886 by Arthur Wright, the engineer of the Brighton supply undertaking, to record the load on the system. He arranged for a strip of paper coated with carbon to be pulled along steadily and marked by a needle on a simple moving iron meter.[8]

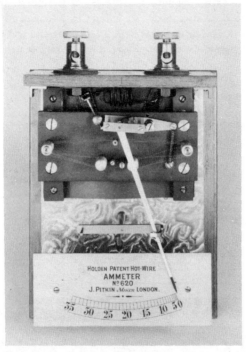

Fig. 12.10   *'Hot-wire' ammeter, in which a taut wire is heated by the current being measured and simultaneously pulled sideways by a spring fixed at its centre*
The deflection is used to move the pointer

In 1883 Crompton and Kapp patented a type of instrument which was designed to overcome the problem of inconstancy in magnetic materials. The moving element was a magnetic needle which was deflected by the magnetic field of a coil carrying the current to be measured and restrained by a constant magnetic field. The constant, controlling magnetic field was obtained by using the same current in another coil to magnetise a very thin iron core to saturation (Fig. 12.11). Both voltmeters and ammeters were developed using this principle. In a typical instrument (Fig. 12.11a) there were four solenoids. One diagonally opposite pair, wound on a nonmagnetic core, provided the deflecting force. The other pair had a core consisting of a piece of iron wire which saturated at a low current and provided the controlling force. The magnetic needle and pointer were free to rotate indefinitely. The magnetic needle was usually astatic, so that the earth's magnetic field did not affect the reading whatever the position of the instrument.

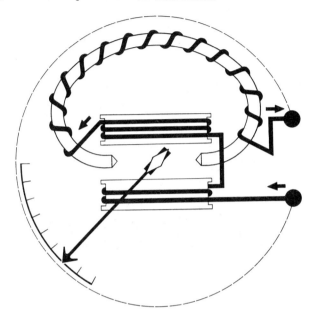

Fig. 12.11a   *Principle of the Crompton-Kapp instruments*
The magnetic needle carrying the pointer can move freely. The deflecting force is provided by the large coils; a constant restraining force is provided by the thin C-shaped iron core which has a coil carrying the same current but is always magnetised to saturation because it is so thin. This instrument does not rely on the stability of magnetic materials, and if the needle is made astatic then it is unaffected by the magnetic field of the earth or of nearby electrical machines

In one form of ammeter intended for power station use (Fig. 12.11*d*) the deflecting force is the magnetic field of a single conductor passing under the needle. Two C-shaped coils with a magnetically saturated thin iron core provide the constant controlling magnetic field. This particular instrument does not have terminals, but a plug composed of two flat conducting plates separated by a sheet of insulating material. This plug could be inserted between a pair of normally touching contacts in a vertical busbar. With this arrangement a single instrument could be used to measure the current in a number of circuits when desired without interrupting the circuit and without the expense of individual instruments for each circuit.

When better magnetic materials became available towards 1900 Crompton abandoned this type of instrument in favour of conventional designs. He also started putting the zero on the left, as other manufacturers did, although the Crompton-Kapp instruments just described have their zero on the right of the scale.

Fig. 12.11*b*   *Crompton-Kapp instrument for 150 A*

Fig. 12.11*c*   *The instrument in (b) with the cover removed.*
The thin iron core is easily seen

Fig. 12.11*d***   *Crompton-Kapp instrument of 100 A***
The 'terminals' are a plug at the top designed to be pressed between two
normally-closed spring contacts in a bus-bar, so that measurements can be made
using a portable instrument without breaking the main circuit

An important practical instrument is Evershed's ohmmeter, the 'Megger',
designed especially for insulation testing (Fig. 12.12). It has a pivoted permanent
magnet and two sets of coils at right angles, arranged so that the position of the
magnet depends on the ratio of the field strengths of the two coils. A battery or
generator passes a current through the circuit being tested. One set of coils carries
the current to the test circuit and the other set of coils is connected in parallel so
as to receive the test voltage. The earlier Meggers were sold with a separate hand-
driven generator, but the instrument and generator are now usually combined in
one case.

12.4   Energy meters

The first supply undertakings based their charges on the number of lamps connected, rather than on power actually used, but there was clearly a need for
instruments that could measure the total energy used. Edison made electrolytic

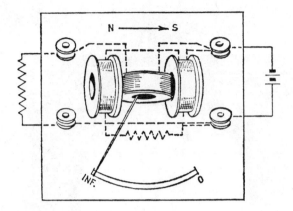

Fig. 12.12   *The Evershed 'Megger' for insulation testing*
A small permanent magnet is suspended at the intersection of the centre lines of
two coils, whose magnetic fields represent the current and voltage in the test
circuit. The magnet takes up a position determined by the ratio of the current and
voltage

meters (Fig. 12.13) in which the current passed through an electrolytic cell consisting of two zinc plates dipped in zinc sulphate solution. The cell could be
shunted by a resistance if large currents were to be measured. It was, strictly, an
ampere-hour meter rather than a watt-hour meter, but on a constant voltage system
the end result is the same and many early supply meters measured ampere-hours.

Fig. 12.13   *Edison's electrolytic house service meter*
The 'meter reader' had to weigh the copper plates in the electrolytic cells

Edison's 'meter readers' weighed the zinc plates and the charge was based on the weight of zinc transferred.

The first electromagnetic supply meter was the clock meter devised originally by Ayrton and Perry in 1882 but improved by Aron and usually known by his name. It depends on the fact that the speed of a pendulum clock varies if 'gravity' varies. Two pendulum clock mechanisms were coupled together through a differential gear to dials which recorded any difference between the readings of the two clocks. One of the pendulums carried a coil of fine wire connected across the supply, and a coil of thick wire in series with the load was placed so that the interaction of the magnetic fields of the two coils varied the apparent force of gravity on the pendulum. The gain or loss of that clock, which was recorded on the dials, was a measure of the power consumed in the circuit. In later versions of the Aron meter both clocks were influenced by voltage and current coils, one being made to gain and the other to lose. The clocks were usually driven by clockwork but rewound automatically.

Most energy meters have what is effectively an electric motor whose driving torque is proportional to the magnetic fields of both 'current' and 'voltage' coils. The motor is restrained by an eddy current brake which gives a retarding torque proportional to speed. The total number of revolutions is then a measure of the energy consumed.

The first motor meter was devised in 1882 by Elihu Thomson (1853-1937), who was born in England but brought up in Philadelphia. He was a student of Edwin James Houston (1847-1914) with whom he formed the Thomson-Houston Company to manufacture arc lighting equipment in 1879. Thomson's meter had two large air-cored coils as field magnets carrying the current, and a coil of fine wire mounted on a vertical axle and connected across the supply rotated within the field coils.

Thomson's meter was intended for use on d.c., although it would work on a.c. also. Ferranti's mercury meter was essentially a d.c. instrument. It had a copper disc, varnished except at its edge, rotating in a shallow bath of mercury between the poles of a pair of permanent magnets. The current being measured flowed radially through a disc, and caused it to revolve in the magnetic field. The number of revolutions was proportional to the quantity of electricity that passed, so this was an ampere-hour meter.

After the invention of the induction motor, induction-type instruments were adopted almost universally for a.c. systems. The Shallenberger meter of 1898 had a light horizontal iron rotor disc and a two-part field coil. Both parts carried the current to be measured, but one part had a short-circuited turn of copper added so that the system acted as a shaded-pole motor. This meter was an ampere-hour meter, but the modern watt-hour meter was developed from it by arranging separate 'voltage' and 'current' coils to create a rotating magnetic field and turn the disc.

12.5 Oscillographs

With alternating current systems there was naturally an interest in the nature of the waveform. Wheatstone's method of obtaining the waveform of an alternating current was described in Chapter 6 and the method was apparently re-invented in 1880 by Joubert, who was studying a.c. arc lighting circuits. It was widely used in electrical laboratories until the 1920s.

A mechanical oscillograph, which would give a visual display of a complete waveform, was suggested in 1892 by Blondel, but the first practical oscillograph was made by Duddell in 1897. William du Bois Duddell studied under Ayrton and took a special interest in measurements. His basic idea was to make a galvanometer whose moving parts were so lightweight that they could follow the variations of an alternating current, and to use a synchronously moving mirror to spread out the indications in time. He used a pair of flat bronze strips carrying a tiny mirror to make a galvanometer movement (Fig. 12.14) whose natural frequency of vibration was 10^{-4}s, far shorter than that of any instrument previously made. Another mirror was driven by a synchronous motor on an axis at right angles to the galvanometer axis. The instrument was described at the British Association meeting in Toronto in 1897, and Duddell said that he could follow cyclic changes occurring in one three-hundredth of a second with his instrument.[11]

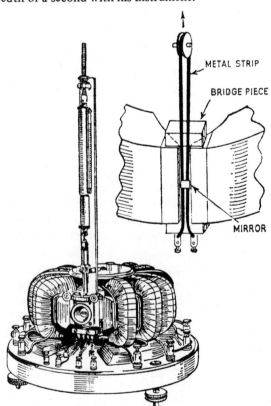

METAL STRIP

BRIDGE PIECE

MIRROR

Fig. 12.14   *Duddell's oscillograph, which was essentially a moving coil galvanometer so light and sensitive it could follow an alternating current waveform*

The discovery of the electron in 1897 and other basic research into electrical phenomena in evacuated glass tubes led to the cathode ray tube. Tubes with either electrostatic deflection or electromagnetic deflection in two axes at right angles were available by about 1900, and used to display waveforms. The first time-base circuits used a capacitor charged through a resistor to give a deflection voltage varying uniformly with time. Cathode ray tube oscilloscopes were developed first for audio frequency investigations connected with the telephone. A review of electrical measuring instruments published by the Institution of Electrical Engineers in 1928 gave much more space to mechanical oscilloscopes than to cathode ray ones,[12] though during the 1930s the cathode ray oscilloscope replaced the mechanical instruments almost completely.

12.6 Electrical standards

The International Congress of Electricians which met in Paris at the time of the International Electrical Exhibition in 1881 was the first body to consider the question of standards on an international basis. Many workers had adopted their own standards of resistance, voltage, etc. and although it was generally agreed that the e.m.f. of a chemical cell (usually a Daniell cell) was a good choice there was little consensus on other units. Wheatstone adopted as his own standard of resistance a piece of copper wire one foot (30·48 cm) long and weighing 100 grains (6 g). He made up resistance boxes, some scaled in miles of wire.

The British Association in 1861 set up a committee to consider the whole question of electrical standards. William Thomson was the first chairman, and in their first report, published in 1862, they said they had decided to adopt a coherent system of units for resistance, current and electromotive force, and to base their units on an electromagnetic rather than electrostatic system. They also decided to relate, where appropriate, to the metric system of weights and measures, rather than to the imperial system, and to use the centimetre, gramme and second as basic units (the 'CGS' system). They decided that a basic standard of resistance should be prepared, equivalent in CGS terms to 10^9 cm/s. The name 'ohmad' was first suggested, but quickly shortened to 'ohm'. Subsequently they developed methods of determining the basic units with ever increasing accuracy. Their procedures included such things as measuring the heat generated by a known current in a standard resistance.

When the Congress met in 1881 many practical engineers were using the 'Weber' as a unit of current, equivalent to about 0·1 A. The Congress recommended the ampere, volt and ohm as practical units, and they were generally adopted by electrical engineers. The new units did not, of course, have any legal force, and there remained also a need for precise definition. In Britain the Board of Trade set up an Electrical Standards Committee in 1891, and this committee sent representatives to another International Electrical Congress which was held at Chicago in 1893. This Congress issued a series of recommendations which were given legal force in Britain in 1894. The units were defined thus:

Resistance: The International Ohm represented the resistance to an unvarying current by a column of mercury at the temperature of melting ice 14·4521 g in mass of a constant cross-section and of a length of 106·3 cm.

Current: The International Ampere represented by the current which when passed through a solution of nitrate of silver in water deposits silver at the rate of 0·001118 of a gramme per second.

Electromotive force: The International Volt which is the e.m.f. represented by 1000/1434 of the e.m.f. of a Clark's cell at 15°C.

The Congress also gave definitions derived from the above for:

The unit of quantity, the International Coulomb, the quantity of electricity involved when a current of one ampere flows for one second.

The International Farad, the capacity whose terminal potential is raised one volt by a charge of one coulomb.

The International Joule, the energy expended when one ampere flows through a resistance of one ohm for one second.

The International Watt, the power when energy is expended at the rate of one joule per second.

The International Henry, the inductance of a circuit in which an e.m.f. of one volt is induced by a rate of change of current of one ampere per second.

With minor changes in definitions these units have been universally accepted and remain in use today. The principal practical change has been that the units are no longer based on the centimetre, gramme, and second (CGS system) but on the metre, kilogramme, and second (MKS system), which has some advantages in theoretical calculations.

The international work on electrical standards is now the responsibility of the International Electrotechnical Commission, formed in 1906 with R.E.B.Crompton as its first secretary. Now based in Geneva, its work extends far beyond standards of measurement. It seeks to promote standards of safety and internationally agreed specifications for electrical equipment to facilitate trade.

12.7 References

1   Anon.: 'A letter from . . . to Mr John Ellicott FRS of weighing the strength of Electrical Effluvia', *Phil. Trans.,* 6 March 1746*, pp.96-99

2   Ellicott, John: 'Several essays towards discovering the laws of electricity', *Phil. Trans.,* 25 February 1748*, pp.195-202; 24 March 1748, pp.203-212; 19 May 1748, pp.213-224

3   Bennet, Abraham: 'Description of a new electrometer', *Phil. Trans.,* 7 and 21 December 1786, **77,** pp.26-34 and plates

4   Coulomb, C.A.de: 'Construction et usage d'une Balance électrique', *Mémoires de l'Académie Royale des Sciences,* Paris 1785, pp.569-611

5   Thomson, W.: 'On new instruments for measuring electrical potentials and capacities', *British Association Report,* 1855, part ii, p.22
6   *Poggendorff's Annalen,* 1841, **54,** p.161
7   Wheatstone, C.: 'An account of several new Instruments and Processes for determining the constants of a Voltaic circuit', *Phil. Trans.,* 1843, **133,** pp.303-327
8   The preceding paragraphs are largely based on references in *The Electrician* during the period
9   British patent no. 4453/1883
10 Joubert, J.: 'Sur les courants alternatifs et la force électromotrice de l'arc électrique', *Journal de Physique,* Paris 1880, **9,** pp.297-303
    (S.P.Thompson, in his *Dynamo-electric machinery,* 7th edn., 1905, **2,** p.234, ascribes the arrangement to Joubert. Thompson was a competent historian, which suggests that Wheatstone's use of the arrangement was forgotten by 1905)
11 *Electrician,* 1897, **39,** p.637
12 Drysdale, C.V.: 'Electrical measuring instruments', *J. IEE,* 1928, **66,** pp.596-616

*Date according to Gregorian calendar: the previous year according to the Julian Calendar in use at the time.

Electrical installations

13.1 Domestic electrical installations

The lampholders, switches, fuses and wiring with which we are familiar today were first developed in the 1880s and have changed little in basic design, although there have been many changes in materials.[1] The modern electrician would look askance at the lack of earthing, the occasional bare live metal, and the fire risk with wooden conduits and accessories.

In Britain the bayonet cap lampholder was introduced in 1884, but Swan's loop holders remained popular for many years. The Edison and Swan United Electric Light Company's catalogue of 1890 offered customers a choice of six terminal arrangements for lamps. There were three alternative loop arrangements, two bayonet caps (one had two contacts as used today, the other had single central contact and used the cap itself for the other connection), and the Edison screw. The bayonet cap was the Company's first choice, described as 'Brass Collar to fit patent Bayonet-joint holder'. It is interesting to note that in 1890 the abbreviation 'BC' in connection with lamps meant 'brass collar', not 'bayonet cap'.[2] The bayonet cap holder has been the most popular holder in Britain ever since, though in most countries the Edison screw is preferred. The 'Goliath' Edison screw for larger sizes of lamp was introduced in 1911.

Early lampholders often incorporated a switch which was sometimes the only switch in early lighting circuits. Many early electrical accessories were made of wood, but porcelain was used extensively from the late 1880s, and the familiar lampholder with brass case and porcelain interior dates from the late 1890s. More recently bakelite and other plastics have been used but the fundamental designs have not changed.

For domestic lighting, as opposed to public lighting, it was essential that individual lamps or small groups of lamps should be switchable independently. Most switches were operated by a turning motion in the 1880s. Edison preferred switches of the turning type because he thought it would be a good selling point if his lights were controlled by the same action as the gas lights he was seeking to replace. The convention was adopted that the handle of the switch should be vertical when 'off' and horizontal when 'on'.

The first manufacturer to specialise in wiring accessories was A P Lundberg & Co. Lundberg had worked with Crompton but left in 1882 to set up his own firm. In 1885 he patented a switch with definite 'on' and 'off' positions, though it did not have a quick-make or quick-break action. Early switches all suffered destructive arcing at the contacts, especially on d.c. systems. Switch contacts had a short life and when, as was often the case, the switch base was made of wood the fire risk was significant.

John Henry Holmes (died 1935) of Newcastle on Tyne was the first to develop and patent a quick-break switch.[3] According to his patent specification of 1884, his switch mechanism 'allows the contact piece or a portion of it by the influence of an independent force, such as that of a spring or weight, to move through a certain distance independently of the handle'. Since the contact moved independently of the handle it could be made to move quickly, even if the handle was only moved slowly. Holmes' first switches had the handle and the moving contact pivoted on a common axis. The handle pressed directly on the moving contact to bring it into engagement with a fixed contact, where it was held by friction, but to open the switch the handle pulled the contact through a spring. The new switches were so superior to the old slow-break designs that sixteen other switch-making firms acquired licences to use Holmes' patent.

Holmes ran a family electrical engineering business in Newcastle, and undertook both electrical installation work and the design and manufacture of electrical machines. His 'Castle' dynamos were used both in England and abroad, and he supplied portable lighting sets to be used on the Suez Canal for navigation at night.

Three years later, in 1887, Holmes patented switches with 'quick-make' as well as 'quick-break', in which the contact piece was moved in both directions by a spring and never directly driven by the operating handle. A selection of Holmes' switches are preserved in the Science Museum at Newcastle upon Tyne (Fig. 13.1).

Fig. 13.1   *Early domestic light switch incorporating quick-break and quick-make actions, designed and made by J.H. Holmes*

Tumbler action switches gradually replaced turn switches after 1890, although the turning action continued to be preferred for heavy-duty switches and in applications where a china cover was wanted. Since there was no provision for earthing the metal covers used on other light switches an insulating cover, usually of china, was preferred in kitchens and other damp situations.

A single push-button switch, which turned 'on' and 'off' in response to alternate pushes, was patented by the Edison & Swan Company in 1893. The single cord pull-switch, popular for bedroom lights, was first made in 1886. A two cord pull-switch, based on an ordinary tumbler switch, was introduced in 1900.

Two-way switches were used at least as early as 1884, in the lighting installation at Peterhouse, Cambridge. The college celebrated its six hundredth anniversary that year and Sir William Thomson (later Lord Kelvin), a former student, gave the college a complete lighting installation. He designed a number of lighting fittings including an adjustable height pendant (Fig. 13.2).

Switches produced for domestic use were often highly ornamented works of art. Falk, Stadelmann & Company's Wholesale Electric Light Accessories catalogue of 1904 offers ten types of wall-mounted light switch, and each type was available in a choice of designs and colours.[4] Retail prices ranged from 3 to 16 shillings (15 to 80 pence). Some switches include terminals for connecting a piece of fuse wire in the circuit. Others were arranged for operating by cords. Two-way and intermediate switches were also available, though the choice of styles was limited. All had china bases and either porcelain or metal covers. The same catalogue includes wooden pendant switches and wooden plugs and sockets. The wood used was cocus, a West Indian wood particularly suitable for turning.

During the 1970s domestic dimmer switches have become popular, especially in living rooms where lights may be required bright for reading or dim for watching television. These switches, which are made as direct replacements for ordinary light switches, use a thyristor arranged to switch the circuit on at a point partway through each half-cycle of the supply. The point at which the thyristor turns on is determined by a simple resistance-capacity circuit, and the lamp is effectively operated on a variable voltage supply. Dimmer switches are ideal for controlling filament lamps, but cannot be used with fluorescent lamps unless special circuits are used that separate the supply to the lamp heaters from the main supply.

In the early days of electricity supply most installations were being put into existing buildings (Fig. 13.3). The main demand, therefore, was for surface wiring and accessories. After the First World War, however, many new houses were being wired during construction. Surface mounted switches gave way to flush switches set into the wall, and to a lesser extent the same applied to sockets. It was necessary to cut switch recesses in the brickwork to accommodate the switches, and much effort went into designing switches shallow enough to be mounted in the depth of the plaster. The first 'plaster-depth' switches were introduced in 1933 by S. O. Bowker Ltd.

Since the Second World War, a.c. supplies have been adopted virtually everywhere. It is no longer necessary to design domestic switches so that they can break the arc

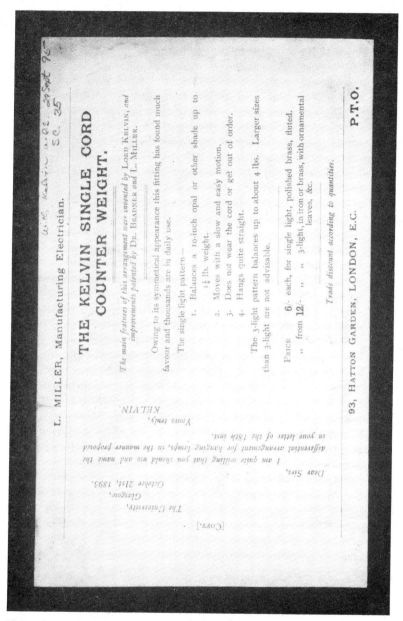

Fig. 13.2 *Advertising leaflet for Kelvin's adjustable light fitting*
(From a leaflet in the IEE Archives)

on a d.c. circuit, and the industry has reverted to simple slow-action switches for most lighting applications.

Plugs and sockets were first used for table lamps, but the variety of domestic

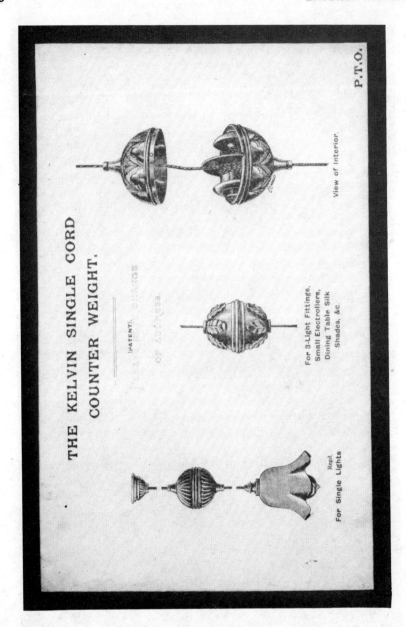

THE KELVIN SINGLE CORD COUNTER WEIGHT.

View of Interior.

For 3-Light Fittings, Small Electroliers, Dining Table Silk Shades, &c.

For Single Lights.

P.T.O.

electrical appliances appearing in the 1890s required plugs and sockets of higher current ratings. Both two pin and concentric contact plugs and sockets were produced by several makers, and a fuse was often incorporated in one or the other, consisting of two terminals holding a short length of fuse wire.

An early development was the shuttered socket, introduced first with the user's safety in mind but also advocated as a floor socket which would not become filled

Fig. 13.3   *Switch and fuse board from a Church in Gloucester about 1900*
Note the bare live metal around the fuses

with dirt. Cromptons designed one in 1893 which had an insulating disc pivoted on the face of the socket (Fig. 13.4). To insert the plug the user had to put the plug pins into two holes in the disc, twist against a spring until the holes in the disc were aligned with the socket contents, and then push the plug fully in. A socket introduced in 1905 had a third pin which operated a shutter; it was thinner than the current-carrying pins and not used as an earth contact.

Interlocking switch sockets, designed to prevent withdrawal of the plug before the switch was turned off, were first patented in 1898 though not much used until 1911. Some switch sockets were designed so that the action of withdrawing the plug also turned the switch off. Such sockets were mainly intended for connecting domestic appliances on d.c. circuits, where arcing could be a problem if an appliance was disconnected by pulling out the plug.

British Standards for two-pin sockets were introduced in 1927, though three-pin sockets were not standardised until 1934. Ratings of 2, 5 and 15 A were chosen. Such sockets were normally supplied 'radially' from a central distribution board, where there was a fuse controlling the wire to each socket (or possibly every two or three sockets).

After the Second World War Britain adopted a new wiring standard, based on a 'ring main' linking sockets all rated at 13 A. Individual appliances use identical plugs but the plugs contain a fuse chosen to suit the appliance. At the time of writing (1981) international discussions are being conducted to find a worldwide

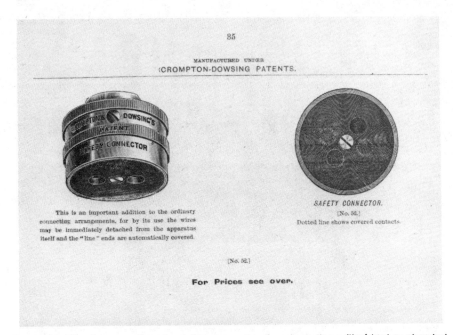

Fig. 13.4   *The Crompton-Dowsing safety connector, for connecting a 'live' lead to electrical apparatus*
The live contact holes are covered by an insulating disc held by a spring. To make a connection the connector is pressed onto the contact pins of the appliance, twisted to expose the live contact holes, and then pushed home

standard plug and socket system.

Insulated wires were being produced for the telegraph industry before they were needed for electric light and power. Insulating materials used were rubber, gutta percha, or textile fibres, such as cotton, impregnated with wax or tar. Rubber had been known since the mid-eighteenth century, and the technique of applying a rubber coating to a metal wire was worked out in the 1830s. In 1839 Charles Goodyear discovered the process of 'vulcanising' rubber, in which the rubber is heated with sulphur yielding a far more durable product. Gutta Percha is a similar substance to rubber, also of vegetable origin. The early telegraph engineers found it preferable to rubber, especially for submarine telegraph cables.

Gutta percha insulated cables were sometimes used for electric lighting, but the material is thermoplastic and it was found that if the cable became warm then the gutta percha 'flowed' and was no longer uniform around the conductor.

For many years vulcanised rubber insulation and an outer lead sheath were the most commonly used for domestic installations. The copper conductors were almost invariably tinned and there was usually a thin layer of pure rubber between the conductor and vulcanised rubber, to prevent chemical action between the copper and sulphur in the vulcanised rubber.

In the early years many sizes and grades of cable were manufactured. For example, one firm alone, W.T. Glover & Co., offered 58 sizes of cable up to one square inch with 14 different grades of insulation in its 1898 catalogue. One of the main objects of the Cable Makers Association, formed in 1900, was to agree upon a range of cable sizes and qualities which gave customers a reasonable range of choice while not requiring an excessive number of types to be stocked. In 1901 there were 41 sizes and 6 grades of rubber insulated cable, and by 1954 the Association had simplified the range to 19 sizes with two alternative thicknesses of rubber, for 250 or 660 V circuits.

Early wiring often used single-core cables enclosed in wooden casing which both protected the cables from mechanical damage and, at least in the earliest installations, ensured the physical separation of conductors when the quality of the rubber insulation was open to doubt.

Wood casing was widely used and gave good service up to the First World War, and in some places later still, but it was very costly since its installation required skilled carpentry. From about the turn of the century it was gradually replaced by metal conduits.

Several manufacturers produced complete wiring systems, with cables, sometimes conduit, junction boxes, switches, sockets etc. Probably the best known and most widely used was the 'Henley Wiring System' introduced in 1911 by W.T. Henley's Telegraph Works Co. Ltd. The Company was a telegraph manufacturer which had expanded its business to embrace electric light and power, but retained its old name. The Henley System used flat twin (or 3-core) cables with rubber insulation and a lead alloy sheath which was harder than pure lead but easily manipulated with the hands. Junction boxes were supplied for making the connections and, when properly installed, they also ensured electrical continuity from the sheath of one cable to the next, so that they were all earthed (Fig. 13.5).

The Henley System's greatest rival was a rubber sheathed cable introduced, also in 1911, by the St Helen's Cable & Rubber Co. Ltd. Formed in 1899, the St Helen's Company manufactured cables and a wide range of other rubber products. It seemed to them that the flexible material used for motor car tyres ought to be tough enough to cover cables, and they introduced their Cab Tyre Sheath, or 'c.t.s.' cables. The individual conductors were insulated first with a layer of pure rubber, then with vulcanised rubber, and the cable of two or three such cores was covered with the special tough cable tyre sheath. The c.t.s. system was viewed with suspicion by some engineers as a retrograde step, but it gave good service and with the shortage of metals in World War I it soon became popular. Other manufacturers produced similar cables known as t.r.s. ('tough rubber sheath'), since c.t.s. was a registered trade mark.

The first synthetic plastic insulated cable was made in 1937, when a German firm introduced p.v.c. insulated cable under the name Mipolim. Since the Second World War, p.v.c. insulation has almost completely replaced rubber, though polythene insulation with a p.v.c. sheath has been used, since polythene has a higher resistivity than p.v.c.

Paper insulation has been used in domestic wiring cables, but only on a small scale. Its use in distribution mains, pioneered by Ferranti, was described in Chapter 9.

An important class of cables is that with minerals, rather than organic materials, for their insulation. Mineral insulated cable was first produced in France in 1934 by the Société Alsacienne des Constructions Mécaniques, and since 1937 it has been made in Britain by Pyrotenax Ltd. The insulating material is magnesia, which is compressed into blocks and then dried. The blocks are fed into a copper tube up to five centimetres in diameter and several metres long. Copper rods are threaded through a hole or holes in the blocks and the whole assembly is put through a drawing and annealing process which is repeated until the cable is reduced to the required size. The cable is robust and needs no conduit or other protection; the joints must be watertight. It is completely fireproof. It is used where the ambient temperature is too high for conventional cables, and also in damp surroundings. A special application is in historic buildings, where mineral insulated cable can be surface mounted unobtrusively and gives a very safe installation.

The need for fuses was recognised as soon as electric lighting progressed beyond the one generator plus one arc lamp system. When Sir William Thomson first installed a generator and lighting at his Glasgow house he thought that a short circuit would demagnetise the generator, making a main fuse unnecessary. Experiment, however, showed that that was not so, and he devised a main 'fuse' consisting of two pointed brass rods joined end to end by a small piece of solder. Springs were arranged to pull the rods apart if the solder melted.

Fig. 13.5   *Display of the Henley wiring system in Croydon Corporation Electricity Showroom, about 1930*
(From a photograph in the South Eastern Electricity Board's Milne Museum, Tonbridge)

Society of Telegraph Engineers and of Electricians.

RULES AND REGULATIONS

FOR THE PREVENTION OF FIRE RISKS ARISING FROM ELECTRIC LIGHTING,

Recommended by the Council in accordance with the Report of the Committee appointed by them on May 11, 1882, to consider the subject.

MEMBERS OF THE COMMITTEE.

Professor W. G. Adams, F.R.S., *Vice-President.*	Professor D. E. Hughes, F.R.S., *Vice-President.*
Sir Charles T. Bright.	W. H. Preece, F.R.S., *Past President.*
T. Russell Crampton.	
R. E. Crompton.	Alexander Siemens.
W. Crookes, F.R.S.	C. E. Spagnoletti, *Vice-President.*
Warren De la Rue, D.C.L., F.R.S.	James N. Shoolbred.
Professor G. C. Foster, F.R.S., *Past President.*	Augustus Stroh.
Edward Graves.	Sir William Thomson, F.R.S., *Past President.*
J. E. H. Gordon.	Lieut.-Colonel C. E. Webber, R.E., *President.*
Dr. J. Hopkinson, F.R.S.	

These rules and regulations are drawn up not only for the guidance and instruction of those who have electric lighting apparatus installed on their premises, but for the reduction to a minimum of those risks of fire which are inherent to every system of artificial illumination.

The chief dangers of every new application of electricity arise mainly from ignorance and inexperience on the part of those who supply and fit up the requisite plant.

The difficulties that beset the electrical engineer are chiefly internal and invisible, and they can only be effectually guarded against by "testing," or probing with electric currents. They depend chiefly on leakage, undue resistance in the conductor, and bad joints, which lead to waste of energy and the production of heat. These defects can only be detected by measuring, by means of special apparatus, the currents that are either ordinarily or for

Fig. 13.6 *The first Wiring Rules, published by the Society of Telegraph Engineers (now the IEE) in 1882*

362   RULES AND REGULATIONS FOR THE PREVENTION OF

the purpose of testing, passed through the circuit. Bare or exposed conductors should always be within visual inspection, since
the accidental falling on to, or the thoughtless placing of other
conducting bodies upon such conductors might lead to "short
circuiting," or the sudden generation of heat due to a powerful
current of electricity in conductors too small to carry it.

It cannot be too strongly urged that amongst the chief enemies
to be guarded against, are the presence of moisture and the use of
" earth " as part of the circuit. Moisture leads to loss of current
and to the destruction of the conductor by electrolytic corrosion,
and the injudicious use of " earth " as a part of the circuit tends to
magnify every other source of difficulty and danger.

The chief element of safety is the employment of skilled and
experienced electricians to supervise the work.

I. THE DYNAMO MACHINE.

1. The dynamo machine should be fixed in a dry place.
2. It should not be exposed to dust or flyings.
3. It should be kept perfectly clean and its bearings well oiled.
4. The insulation of its coils and conductors should be perfect.
5. It is better, when practicable, to fix it on an insulating bed.
6. All conductors in the Dynamo Room should be firmly supported, well insulated, conveniently arranged for inspection, and
marked or numbered.

II. THE WIRES.

7. Every switch or commutator used for turning the current on
or off should be constructed so that when it is moved and left to
itself it cannot permit of a permanent arc or of heating, and its
stand should be made of slate, stoneware, or some other incombustible substance.

8. There should be in connection with the main circuit a safety
fuse constructed of easily fusible metal which would be melted if
the current attain any undue magnitude, and would thus cause the
circuit to be broken.

9. Every part of the circuit should be so determined, that the
gauge of wire to be used is properly proportioned to the currents
it will have to carry, and changes of circuit from a larger to a
smaller conductor, should be sufficiently protected with suitable

FIRE RISKS ARISING FROM ELECTRIC LIGHTING.   363

safety fuses so that no portion of the conductor should ever be allowed to attain a temperature exceeding 150° F.

N.B.—These fuses are of the very essence of safety. They should always be enclosed in incombustible cases. Even if wires become perceptibly warmed by the ordinary current, it is a proof that they are too small for the work they have to do, and that they ought to be replaced by larger wires.

10. Under ordinary circumstances complete metallic circuits should be used, and the employment of gas or water pipes as conductors for the purpose of completing the circuit, should in no case be allowed.

11. Where bare wire out of doors rests on insulating supports it should be coated with insulating material, such as india-rubber tape or tube, for at least two feet on each side of the support.

12. Bare wires passing over the tops of houses should never be less than seven feet clear of any part of the roof, and they should invariably be high enough, when crossing thoroughfares, to allow fire escapes to pass under them.

13. It is most essential that the joints should be electrically and mechanically perfect. One of the best joints is that shown in the annexed sketches. The joint is whipped around with small wire, and the whole mechanically united by solder.

14. The position of wires when underground should be efficiently indicated, and they should be laid down so as to be easily inspected and repaired.

15. All wires used for indoor purposes should be efficiently insulated.

16. When these wires pass through roofs, floors, walls, or partitions, or where they cross or are liable to touch metallic masses, like iron girders or pipes, they should be thoroughly protected from abrasion with each other, or with the metallic masses, by suitable additional covering; and where they are liable to abrasion

from any cause, or to the depredations of rats or mice, they should be efficiently encased in some hard material.

17. Where wires are put out of sight, as beneath flooring, they should be thoroughly protected from mechanical injury, and their position should be indicated.

N.B.—The value of frequently testing the wires cannot be too strongly urged. It is an operation, skill in which is easily acquired and applied. The escape of electricity cannot be detected by the sense of smell, as can gas, but it can be detected by apparatus far more certain and delicate. Leakage not only means waste, but in the presence of moisture it means destruction of the conductor and its insulating covering, by electric action.

III. LAMPS.

18. Arc lamps should always be guarded by proper lanterns to prevent danger from falling incandescent pieces of carbon, and from ascending sparks. Their globes should be protected with wire netting.

19. The lanterns, and all parts which are to be handled, should be insulated from the circuit.

IV. DANGER TO PERSON.

20. To secure persons from danger inside buildings, it is essential so to arrange the conductors and fittings, that no one can be exposed to the shocks of alternating currents exceeding 60 volts; and that there should never be a difference of potential of more than 200 volts between any two points in the same room.

21. If the difference of potential within any house exceeds 200 volts, whether the source of electricity be external or internal, the house should be provided outside with a " switch," so arranged that the supply of electricity can be at once cut off.

By Order of the Council.

F. H. WEBB, *Secretary.*

Offices of the Society,
4, The Sanctuary, Westminster,
June 21, 1882.

The installation Thomson provided at Peterhouse had fuses for each branch circuit. They consisted of two pieces of copper wire soldered together, hung across two grooved terminals and kept taut by a lead weight at each end. This assembly was mounted in a wooden box with a glass front, and in some cases the boxes were lined with asbestos.

The idea of using a weight to assist rupture of the fuse was taken up by A.C. Cockburn who patented the 'Cockburn Cut-out' in 1887. This was a fuse of tin wire, supplied with terminal loops and a lead weight fixed in the middle.

Tin, or a low melting point alloy, was used for the fusible link rather than copper in order to minimise the fire risk when the fuse blew. In 1889 the firm of Laurence, Paris & Scott designed a safe high melting point fuse using a copper or silver wire in a china tube. Much thought was given to the problem of containing and quenching the arc when a fuse blew. One solution was the cartridge fuses in which the fusible link was totally enclosed within a refractory tube to avoid any fire risk. Cartridge fuses were made by several manufacturers in the 1890s though not widely used. High capacity cartridge fuses became common in the 1920s; small cartridge fuses for fused plugs and clock connectors only came into domestic use with the 13 A ring main system after the Second World War.

It is interesting to note that in early lighting installations, in which the voltage regulation was not very good, fuses were intended to protect the lamps against overvoltages as well as to protect the generator and wiring against excessive currents due to faults. Because the resistance of a carbon filament lamp falls with increasing temperature, a voltage surge produces a disproportionate rise in current. It is therefore possible to use a fuse to protect a carbon lamp against overvoltage, but a metal filament lamp, which has the opposite temperature characteristic, cannot be protected in this way.

13.2 Safety: wiring regulations

Although quite high voltages were used in series arc lighting systems, the early indoor installations rarely exceeded 100 V. Risk from electric shock was not therefore a problem, and indeed some early comments sound very foolhardy when read today. In his *Reminiscences* Crompton recalls[5]

> The Prince of Wales, afterwards Edward VII, and Princess Alexandra visited our exhibition more than once, and I personally had the honour of conducting her Royal Highness round, to point out and explain to her the electrical phenomena. I assured her that she could touch bare conductors carrying energy of 200 volts pressure without danger. She made the experiment, and agreed with me that such a pressure was quite safe to use.

The risk people did appreciate was that from fire, and the earliest wiring rules were drawn up by, or on behalf of, fire insurance companies. The consulting engineer Killingworth Hedges published in 1886 a book *Precautions on introducing the electric light*, intended for owners of large houses who were installing generating plant.[6] He offered a set of instructions which 'should be hung up in the engine-room and should be read over and explained to the man in charge'. These were operating instructions rather than installation instructions, and include the quaint injunction 'Never cool a heated dynamo with water'.

Long before it became the Institution of Electrical Engineers, the Society of Telegraph Engineers published 'Rules and Regulations for the prevention of fire risks arising from electric lighting'.[7] Their first rules, issued in 1882, were a mixture of instructions and good advice. They are reproduced in full as Fig. 13.6. Surprisingly they are solely concerned with arc lighting: there is no mention of filament lamps. The Institution has revised and extended the Rules from time to time, the fifteenth edition being published in 1981. They have rarely had any legal force, but are generally accepted as the guide to sound practice in electrical installations.

13.3. References

1   For a study of wiring and accessories generally see Mellanby, John: *The history of electric wiring*, Macdonald, 1957

2   Edison & Swan United Electric Light Co. Ltd., *Catalogue*, 1 October 1890, p. 12

3   The Holmes' switch collection and biographical notes about Holmes in the Museum of Science and Engineering, Newcastle upon Tyne. Holmes' patents were nos. 3256 of 1884 and 5648 of 1887

4   Falk, Stedelmann & Co. Ltd., *Catalogue No. 195: 'Electric light accessories'*, London, 1904

5   Crompton, R.E.B.: *Reminiscences*, Constable 1928, p.100

6   Hedges, Killingworth: *Precautions on introducing the electric light,* Spon, 1886, pp.47-49

7   *J. Soc. Tel. Eng.,* 1882, pp. 361-364

Chapter 14

Electric lighting

14.1 Metal filament lamps

By the end of the 19th century the carbon filament lamp was in widespread use but a serious competitor had emerged. In the early 1880s 'gas-light' had meant simply the light of a naked gas flame, and the superiority of the carbon filament lamp was indisputable. Between 1885 and 1893 the Austrian Carl Auer von Welsbach developed the first practical gas mantle, which revolutionised gas lighting. Gas light became cheaper than electric light, especially in Britain where gas was relatively cheaper then in most other countries, including the USA.[1]

The mantle utilised the property, possessed by certain metal oxides, of selectively emitting radiation in the visible range when heated. The principle had been used since the mid nineteenth century in theatres for the 'lime-light', in which a block of lime (calcium oxide) was heated by a gas flame. Von Welsbach found that the best mantle was made by impregnating a silk or cotton fabric with a mixture of 99% thorium oxide and 1% cerium oxide. His first mantles were held above the flame by a fireclay support, but by 1903 'inverted' mantle burners were developed in which the mantle hung downwards and the gas flame was fitted to the mantle shape.

The Nernst lamp produced light by electrically heating materials similar to those used in gas mantles. The oxides used are insulators when cold but become conducting when heated. Nernst's idea was to heat a rod of oxide and then keep it hot by passing a current through it. An auxiliary heater was needed for starting and there had to be some arrangement to limit the current since the rod alone would have a negative resistance. Professor Walther Nernst (1864-1941), a German physicist, published the principle in 1898 and Nernst lamps were commercially available in 1900. The first lamps were started with the aid of a small spirit burner, but electrically heated starters were soon introduced that cut out automatically when the oxide rod became conducting. The current was limited by a series resistance of iron wire. Although dearer in first *cost* and more complicated than a filament lamp, the Nernst lamp had several advantages. Its efficiency was about eight lumens per watt, allowing for the loss in the series resistance. The lamp worked in air, although it was provided with a removable glass cover. When the rod failed, it could be replaced cheaply and easily. The lamp was less sensitive to voltage fluctuations and could be made for higher voltages than a filament lamp.[2]

In the search for an improved filament lamp three metals seemed promising: osmium, tantalum and tungsten. All are brittle metals, extremely difficult to draw out into fine filaments. Osmium filament lamps were on sale in 1899, closely followed by tantalum. Their greater efficiency gave them an economic advantage over carbon, despite their relatively high initial cost and the fact that they were so brittle. They were made mainly in Germany, and soon superseded by tungsten lamps.

Tungsten has the highest melting point of all metals, but it could not be drawn into a fine filament until William Coolidge developed a process for doing so in 1909. Some tungsten lamps had been made before that by mixing tungsten powder with a binder, extruding through a die, and then heating to drive off the binder and sinter the powder. After much research Coolidge produced ductile tungsten by compressing tungsten powder, then heating and hammering it. Drawn tungsten filaments quickly replaced the other metal types, although the mass production of carbon filament lamps continued into the 1920s.

In an evacuated tungsten lamp metal vaporises from the filament and is deposited on the glass, causing blackening. The upper limit to the temperature at which a vacuum tungsten lamp could be run was determined by the acceptable rate of blackening. It was usually economic to replace a lamp before the filament failed. A possible improvement was to fill the bulb with an inert gas, but the gas cooled the filament.

Irving Langmuir made a theoretical study of the cooling of a hot wire when gas is present. He found that the rate of heat loss is dependent on the length of the wire but is not much affected by the diameter of the wire. Consequently, if a lamp filament is wound into a tight coil it behaves like a much shorter filament and the heat loss is greatly reduced. The first gas-filled, coiled-filament lamps were filled with nitrogen and were on the market in 1913. Subsequently a mixture of 90% argon and 10% nitrogen was adopted. Because argon has a greater density and lower thermal conductivity than nitrogen, its use decreases both the rate of evaporation and the heat loss from the filament.

The drawn tungsten filament in a vacuum gave a lamp efficiency of about 9 1m/W (lumens per watt). The gas-filled lamp with a coiled filament had an efficiency of about 11 1m/W, and the modern 'coiled-coil' lamp, in which the coiled filament is itself coiled, has an efficiency of about 13 1m/W. The coiled-coil filament was introduced for general purpose lamps in 1934, though had been used earlier in lamps for special applications such as projectors.[3]

Until the early 1920s lamps were usually evacuated through the end away from the cap, leaving a characteristic 'pip' where the bulb was sealed. The pip was easily broken, and lamps are now evacuated through a tube at the cap end so that the fragile point of closure is protected by the cap.

Many special designs of filament lamp have been developed for special applications, such as in projectors, and for stage lighting and flood lighting. Low voltage lamps are made for road vehicles, including the 'sealed beam' type first introduced in America in 1938. In these two thick glass pressings are used. A reflective layer of aluminium is evaporated onto one of the pressings and then the filament is located precisely in relation to the reflector before the pressings are sealed together, evacuated, and filled with gas.

Tungsten halogen lamps became a practical possibility about 1959. In these the bulb contains a halogen, usually iodine or bromine. The halogen reacts with any tungsten which evaporates and drifts to the bulb to form a gaseous tungsten halide which breaks down at the hotter filament. The result is that metal which evaporates from the filament is quickly returned to it, and the lamp can be operated at a higher temperature and therefore greater efficiency when the halogen is present.

It has been recognised almost from the beginning that the life and efficiency of a filament lamp are interchangeable. A high operating temperature results in high efficiency and short life. At lower temperatures the life becomes almost indefinite but the efficiency falls dramatically. Professor John Perry, writing in 1888 about tests made in 1881, said:

At that time we obtained very much higher electrical efficiency from lamps than is ever obtained now from lamps in regular use because at that time it was not sufficiently well known that lamps may be made to give out too much light for their lives.

The life of carbon filament lamps seems to have been typically 400 to 2000 hours, with 1000 hours regarded as normal. In 1926 the Institution of Electrical Engineers discussed a paper 'The economics of lamp choice' by D.J. Bolton, in which the author showed how the optimum operating conditions for a lamp could be determined from considerations of first cost, efficiency, and the cost of electricity. Bolton concluded that lamps of different wattages should be rated for different lengths of life. Higher powered lamps should have shorter, but more efficient, lives than lower powered ones. An example he gave was that a 40 W lamp costing 2s (10p) was correctly given a life of 1000 hours if electricity cost 3d (1·67p) a unit, but the most economical life of a 60 W lamp costing 2s 3d (11·25p) was 882 hours, and the most economical life of a 20W lamp costing 2s 6d (12·5p) was 2230 hours.

For general service tungsten filament lamps most manufacturers have claimed an average life of 1000 hours. The life and efficiency of both filament and fluorescent lamps was the subject of a detailed enquiry by the House of Commons Select Committee on Science & Technology in 1977 and 1978. The enquiry was made in response to allegations that manufacturers designed their lamps to have an unreasonably short life, so as to boost sales, but the Committee concluded that there was no truth in the suggestions and that the combination of life and efficiency adopted by the manufacturers was in the public interest.[4]

14.2   Discharge and fluorescent lamps

Although the fact that an electric discharge in rarefied gas gave a light was noted in the 18th century, it was not until nearly 1900 that the phenomenon was exploited practically. The Moore tube, introduced in 1895, was a carbon dioxide filled tube up to sixty metres long, which gave a white light. The efficiency, about two 1m/W, was too low for general illumination, but from about 1905 the tubes were shaped into letters for advertising purposes. Nitrogen-filled tubes, which give a pink light, were introduced about the same time, and the familiar red neon lights were first made in 1910 (Fig. 14.1). However, it was not until the 1920s that such lights were widely used.

Fig. 14.1   *The West End Cinema, London, in 1913*
The arch above the door is outlined in Moore tubes and the name is in neon tubes.
Arc lamps can also be seen outside the shop windows

The first low pressure mercury arc lamp, developed by Peter Cooper-Hewitt in 1901, consisted of a glass tube 1 m long and 3 cm in diameter. One end was enlarged and contained a pool of mercury which formed one electrode, and the discharge was struck between the mercury pool and an iron electrode at the opposite end of the tube. The first Cooper-Hewitt lamps were started by tilting the tube by hand, but automatic starters were soon made, incorporating an electro-magnet to tilt the tube.

The Cooper-Hewitt lamp was never widely adopted, possibly because its poor colour rendering made it unattractive and its efficiency was not much higher than that of a tungsten filament lamp. However, measurements of the light emitted by mercury discharges revealed that the pressure-efficiency characteristic had a second, higher peak of efficiency at an operating pressure greater than that used by Cooper-Hewitt. This led in the early 1930s to the high pressure mercury vapour lamp, operating at about atmospheric pressure (Fig. 14.2). A major practical problem was to find a glass that would seal to tungsten or molybdenum and operate at 500°C.[5]

Sodium vapour lamps were developed about the same time as high pressure mercury lamps (Fig. 14.3). Sodium vapour reacts with most ordinary glasses, and it was necessary to develop special types of glass for sodium lamps. To reduce heat loss, the lamp itself was usually placed inside a glass vacuum flask. In early lamps the flask was detachable, although a one-piece construction is now used.

Fig. 14.2 *Early high pressure mercury vapour lamp*
The mercury discharge is contained within the inner glass tube.

The low pressure sodium vapour lamp was developed by Philips in Holland. The Philips Gloeilampenfabrieken Company had been founded by the brothers Gerard and Anton Philips in 1894 to make incandescent filament lamps, and that was their main product until they began to manufacture radio valves in the early 1920s. They developed the sodium lamp for street lighting, where its high efficiency was desirable and the fact that colours could not be distinguished in its monochromatic yellow light was not important. The first installation of sodium street lights was at Limburg, Holland, in 1932, and they were used on Purley Way, Croydon, soon afterwards.

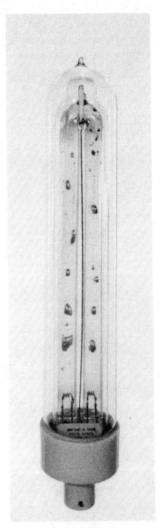

Fig. 14.3 *Early low pressure sodium vapour lamp*
The discharge takes place in the inner 'U' tube, which can be withdrawn from the surrounding glass envelope

Further research has led to the development of glasses that withstand sodium vapour at higher temperatures and pressures than used to be possible. During the late 1970s high pressure sodium lamps have largely replaced low pressure ones, and virtually all new lighting schemes for major roads use high pressure sodium. The advantage is that when under pressure the sodium discharge gives light over a broad band of the spectrum, and although the light has a yellow-pink tinge the colour rendering ability is fairly good.

The fluorescent lamp, introduced about 1940, uses a low pressure mercury discharge which emits most of its radiation in the ultra-violet. The fact that some materials had the property of converting ultra-violet to visible light had been known for a century, and various improved 'phosphors' were produced in the late 1930s. A major breakthrough came in 1942 with the invention in Britain of halophosphate phosphors, with a conversion efficiency between two and three times that of the best materials previously used. A typical fluorescent lamp of 1940 had an efficiency of 23 1m/W. Many fluorescent lamps today have efficiencies of over 60 1m/W. Several different phosphors are in commercial use today, some giving higher conversion ratios and others giving better colour rendering. The choice of phosphors for fluorescent lamps depends on the relative importance of colour and efficiency in the particular application.[6]

A very recent development at the time of writing is the fluorescent lamp with the necessary control gear built into the bulb. Fluorescent lamps have not been much used domestically, despite the fact that the overall cost of light from a fluorescent lamp is about half that from a tungsten filament lamp. Part of the reason why fluorescent lamps are not used more is undoubtedly people's dislike of their colour and the flicker often associated with them. Modern fluorescent lamps are better in both these respects than earlier ones, but, since fluorescent lamps (like all discharge lamps) need a 'ballast' inductance to limit the current they draw, they cannot simply be plugged in to the fittings provided for filament lamps. In 1981 several fluorescent lamp manufacturers have introduced lamps with the ballast incorporated in the bulb; these can be used as direct replacements for filament lamps. Whether these prove a commercial success remains to be seen.

14.3   References

1   Byatt, I.C.R.: *The British Electrical Industry 1857-1914*, Clarendon Press, Oxford, 1979, p.3
2   Details of the Nernst lamp and its introduction are given in numerous short articles in the *Electrician*, 1899, 1900
3   Ruff, H.R.: 'Light sources', *Light and Lighting*, January 1958, pp. 6-13
4   Third report from the Select Committee on Science and Technology, Session 1977-78, especially Appendix 22, *Life and light output of filament lamps*, a Memorandum by Brian Bowers and C.N.Brown
5   Ruff, H.R.: *op. cit.*, pp.6-13
6   Willoughby, A.H.: 'The evolution of electric lamps', *Lighting Research & Technology*, 1969, **1**, pp.69-77

Chapter 15

Electric heating

15.1 Off-peak electricity

At first the public supply system was used exclusively for lighting, but this led to an inefficient use of plant. The demand was concentrated in a few hours each day, from dusk until bedtime; during the remainder of the 24 hours expensive generating plant and mains were standing idle, or, at best, very little used.

With an alternating current system at least one generator had to be running at all times if the supply was to be maintained. A major advantage of direct current systems was that battery storage could maintain the supply during slack periods and the generators could be run at full load, and therefore maximum efficiency, for long enough to supply the peak load and recharge the batteries.

In 1888 Crompton gave the Society of Telegraph Engineers a detailed analysis of the capital and running costs of a.c. and d.c. systems.[1] He chose a 10 000 light or 600 kW plant supplying 1000 houses, which he regarded as an economic size of installation. Both systems used high voltage distribution. The a.c. system had a transformer for every pair of houses to step down the voltage. The d.c. system had four battery substations used as described in Chapter 9 to reduce the voltage to the houses.

Crompton's detailed costing (Table 15.1) shows that the capital cost of the d.c. system was higher than that of the a.c. one, because the batteries cost very much more than the extra generating plant required for a.c. The running costs for d.c., however, were much less than for a.c. Table 15.2 is Crompton's analysis and shows that the d.c. system is cheaper, partly because the a.c. plant needed to be manned 24 hours a day, but mainly because the a.c. plant needed 70% more coal than the d.c. for the same output.

The calculations are based on electricity sales of 2100 kWh per day, whereas if the plant had run at full load all day the output would have been 600 x 24 = 14 400 kWh. The ratio of these (the 'load factor') was 15%. It is clear from these figures that if a demand for electricity could be found during the hours when lighting was not required then that demand could be supplied without building new plant, and

Table 15.1 *Cost of 10000 light, or 600 kW, plant*

A.T. : Alternating transformer distribution

	£
Generating station, buildings, chimney shaft, water tanks, and general fittings	11 000
Dynamos and exciters — 865 kW, including spare sets, divided as convenient	5 540
Motive power, i.e. engines, boilers, steam and feed connections, belts, etc, at £8 12s per i.h.p.	12 470
500 transformers, i.e. one to every pair of houses, at £15 each	7 500
2 000 yards primary or charging main, exterior to area of supply, at £308 per 100 yards	6 160
20 000 yards distributing main, 50 m/m sectional area, at £91 7s	14 270
Regulating gear	500
	£57 440

B.T. : Accumulator transformer distribution

	£
Generating station, buildings, chimney stack, water tanks, and general fittings	8 000
Dynamos — 600 kW, in 6 sets of 100 kW each	4 800
Motive power, i.e. engines, boilers, steam and feed connections, etc. at £8 12s per i.h.p.	8 600
4 groups of accumulators, in all 240 cells, in series, at £40 per cell, including Stands	9 600
2 000 yards charging main, at £306 17s 6d per 100 yards ...	6 137
20 000 yards distributing main, 161·25 m/m sectional area, at £100 12s 6d	20 125
Regulating gear	2 500
	£59 762

Table 15.2 *Working expenses and maintenance of 10 000-light, or 600-kW, plant*

A.T. : Alternating transformer system

	£	s.	£	s.
Materials				
Coals: 4 380 tons at 17/-	3 723	0		
Oil, water, and petty stores: 1 500 hours at 7/6 + 7 250 hours at 1/-	925	0		
Total cost of material			4 648	0
Labour				
2 foreman drivers at 45/-; 6 drivers at 30/-; 9 firemen at 24/-; sundry labour	1 388	8		
Salaries				
1 chief at £500; 2 assistants at £200 each; 4 clerks at £80 each	1 220	0		
			2 608	8
Maintenance of plant				
Motive-power and dynamos: 10% on £18 010	1 801	0		
Building and fittings: 5% on £11 000	550	0		
Transformers: 10% on £7 500	750	0		
Mains: 7½% on £20 430	1 532	5		
Regulating gear: 10% on £500	50	0		
			4 683	5
			11 939	13

2 100 units × 365 days
= 766 500 units.
Cost per unit   3·75d

B.T. : Accumulator transformer system

	£	s.	£	s.
Materials				
Coals: 2 550 tons at 17/-	2 167	0		
Oil, water, and petty stores: 1 400 hours at 5/-	350	0		
Total cost of material			2 517	0
Labour				
1 foreman driver at 45/-; 2 drivers at 30/-; 3 firemen at 24/-; sundry labour	975	0		
Salaries				
1 chief at £500; 1 assistant at £200; 4 clerks at £80 each	1 020	0		
			1 995	0
Maintenance of plant				
Motive-power and dynamos: 10% on £13 400	1 340	0		
Buildings and fittings: 5% on £8 000	400	0		
Accumulators: 15% on £9 600	1 440	0		
Mains: 2½% on £26 262	656	10		
Regulating Gear: 10% on £2 500	250	0		
			4 086	10
			8 598	10

Cost per unit   2·7d

COOKING BY ELECTRICITY.

ABSOLUTELY THE BEST SYSTEM OF COOKING.

MATIEU WILLIAMS, whose works on Cookery have become classic, stated :

" The ideal cooking arrangement is an oven which will **RADIATE HEAT FROM ALL DIRECTIONS UPON THE FOOD TO BE COOKED.**"

THIS IDEAL HAS NOW BEEN REALISED

by the application of Electricity to Cooking. The **ELECTRIC OVEN** is the **ONLY** one that fulfils these conditions.

IT IS NOT AN EXPERIMENT.

IT IS A PRACTICAL SUCCESS.

Electric Cooking and Warming Apparatus of the different descriptions given below can be seen in

ACTUAL PRACTICAL WORK

AT 92 & 93, QUEEN STREET, CHEAPSIDE, E.C.,

where they are being exhibited by THE CITY OF LONDON ELECTRIC LIGHTING COMPANY, LIMITED, from whom any apparatus can be purchased or hired and every information obtained.

Electric Cooking and Heating Apparatus can be made to work at any desired voltage to suit private installations; prices on application.

OVENS.	HOT PLATES.	COFFEE MACHINES
HOT CUPBOARDS.	KETTLES.	SAUCEPANS.
GRILLS.	FISH KETTLES.	RADIATORS.
FOOTWARMERS.	CIGAR LIGHTERS.	FLAT IRONS.

CURLING-TONGS HEATERS.   MOTORS, &c.   &c.   &c.

July. 1894.

Fig. 15.1   *Leaflet published by the City of London Electric Lighting Co. in 1894, offering appliances for sale or hire*

Electric Cigar Lighter.
(For Table)

Price from   **£1 1 0**

Also arranged as a Bracket to fix
on wall.

Electric Hot-plate.
(Rectangular.)

	Price.	Hire per Quarter.
9½" × 11½"	£1 10 0	2/-
12" × 16"	2 2 0	2 6

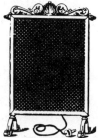

Radiator.
Made in wrot. or cast iron.
Round, 6" diam.. **£2 10 0**
Square, from   **3 10 0**
Hire, **3/6** per Quarter.

Saucepan.

| ¾ pint | £1 10 0 |
| 4 ,, | 2 2 0 |

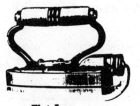

Flat Iron.
| 7 lbs. | £1 10 0 |
| 14 ,, | 1 15 0 |

Hat Irons.
| 8 lbs. | £1 15 0 |

Billiard Irons.
| 20 lbs. | £2 5 0 |

Electric Kettle.
1 pint	£1 15 0
2 ,,	2 0 0
3 ,,	2 10 0
6 ,,	3 10 0
Hire, **3/6** per Quarter.

Electric Ove

Size	Price
21" × 12" × 12"	£7 10
24" × 13½" × 13½"	8 10
24" × 15" × 15"	10 10
24" × 18" × 18"	14 14

Electric Hot-plate.
(CIRCULAR.)

6" diam., wrot. iron stand	£1	1	0
6" .. nickeled ..	1	10	0
8"	1	15	0

Electric Coffee Urn.

2 pint	£5 15 0
4 ..	6 15 0

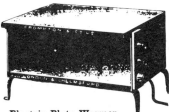

Electric Plate Warmer.

	Price.	Hire per Quarter.
23" × 15½"	£2 10 0	2 6
23" × 17½"	3 3 0	3 6

Hot-plate with Handle.
£1 10 0

Tailor's Iron.

16 lbs.	£1 18 0
21 ..	2 2 0

Electric Saucepan.

	Single Circuit, Fast.	Double Circuit, Fast and Slow.
2 pint	£1 17 6	—
3 ..	2 5 0	—
4 ..	—	£2 10 0
6 ..	3 5 0	3 15 0
8 ..	—	4 15 0

Electric Kettle.

𝕿𝖍𝖊 𝕮𝖎𝖙𝖞 𝖔𝖋 𝕷𝖔𝖓𝖉𝖔𝖓 𝕰𝖑𝖊𝖈𝖙𝖗𝖎𝖈 𝕷𝖎𝖌𝖍𝖙𝖎𝖓𝖌
𝕮𝖔𝖒𝖕𝖆𝖓𝖞. 𝕷𝖎𝖒𝖎𝖙𝖊𝖉.

HEATING and COOKING by ELECTRICITY

THE CITY OF LONDON ELECTRIC LIGHTING COMPANY, LIMITED, are now in a position to SELL or to LET on HIRE ELECTRIC COOKING and WARMING APPLIANCES, and to supply Electricity for Cooking, Warming and Motive Power at

4d. per BOARD OF TRADE Unit.

Electricity has enormous advantages when applied to Heating and Cooking. For example, it is

SAFER THAN GAS.   SIMPLER THAN GAS.
MORE ECONOMICAL THAN GAS.
BETTER THAN GAS IN EVERY WAY.

No danger; no combustion; no chimney; no fires to light and watch; no matches to look for; no heat wasted; perfect regulation. No more burnt meats or pastry; no explosions; pure Air; cool Kitchens.

THE ONLY SANITARY METHOD of PREPARING FOOD.

ENTIRE ABSENCE OF
FIRE. SMOKE. SMELL. DIRT.
AND MOST IMPORTANT OF ALL
NO POISONOUS GASES AROUND FOOD WHILE COOKING.

INVALUABLE IN RESTAURANTS, LUNCHEON BARS AND REFRESHMENT BUFFETS.

FOR PRICES AND HIRE RATES SEE OTHER SIDE.

probably with no increase in staff. Two such loads were potentially available, and were actively sought by the supply undertakers, who offered special cheap tariffs for 'off-peak' electricity. One possible load was electric traction, and many of the early electric tramways were run by or operated in conjunction with the supply undertakings. The other possible load was electric heating and cooking.

15.2   Domestic heating and cooking

Many supply undertakers offered reduced price electricity for electric cookers, and sometimes for electric heaters too. To give further encouragement they offered a range of cookers and other equipment on hire. A City of London Electric Lighting Company leaflet of 1894 (Fig. 15.1) offered four sizes of cooker, which could be purchased for £7. 10s to £14. 14s or hired for 7s (35p) to 12s (60p) per quarter. Kettles, hotplates, plate-warmers, and radiators were available on similar terms. Cheaper items such as electric irons and coffee urns are advertised in the same leaflet but appear to be available for outright purchase only.

Electric cooking demonstrations were arranged regularly. A School of Electric Cookery, under the management of a Miss Fairclough, was fitted up by Crompton in Gloucester Road, London, in 1894. The various cooking appliances, mainly saucepans with built-in elements, stood in a line at the back of the cook's wooden work-table, and each plugged in to a socket on the table.

Although it was well known that a wire could be made hot by passing a suitable current it was not easy at first to make satisfactory heating elements. The reason was that all known metals and alloys oxidised when heated in air. The first solution was to bury the heating wire, which was usually of iron, in the middle layer of three layers of enamel on a cast iron base. Crompton's made heating elements in this way, and by 1900 they were advertising (Fig. 15.2) a range of ovens, hot-plates, hot cupboards (some with polished wooden outer cabinets for use in the dining room), saucepans with built-in elements, several types of iron, kettles, radiators and curling tong heaters.[2]

Crompton's collaborators in the design and manufacture of electric cooking equipment were H. Dowsing and E.J. Fox. They conducted a vigorous advertising campaign in the 1890s. Dowsing wrote booklets and articles expounding the virtues of electric cooking. One fact which strikes the reader of their articles almost a century later is that they felt it necessary to convince their readers that electric cooking was *possible*, let alone preferable. They next concentrated their arguments on the question of cost. They pointed out that an electric oven could be totally enclosed and waste virtually no heat, whereas coal and gas ovens inevitably wasted most of the heat produced. They published figures showing the cost of boiling water, grilling a chop, etc., by various methods. They quoted Matieu Williams, a writer on cookery, who stated 'the ideal cooking arrangement is an oven which will radiate heat from all directions upon the food to be cooked'. Crompton's ovens had heating plates on all sides and at the top and bottom, all separately controlled. It was then possible, for example, to cook a pudding uniformly using all the heating plates and then to brown the top of the pudding by using only the heating plate at the top. The oven could have a uniform temperature throughout, whereas gas ovens could vary 30°C in temperature between the top and the bottom. The dangers of gas were vividly pointed out.[3]

Fig. 15.2   *Pages from Crompton domestic appliances catalogues, about 1896*

Crompton and Dowsing designed electric heaters using carbon filament lamps as the heating element. The lamps were run at lower temperature than when used for lighting, so they gave a red glow as well as heat and have a very long life. Usually the lamps were rated at 250 W and coloured red or orange. Two or four lamps would be mounted together with a reflector panel, so that 500 or 1000 W of heat was radiated forwards (Fig. 15.3).

Several electrical appliance firms were founded before the First World War. Often the founders were people who had not previously been connected with electric lighting or with the supply industry. The General Electric Apparatus Company began as an electrical warehouse in London and was not at first a manufacturer. As the demand for appliances grew, however, it established a factory in Birmingham and in 1889 changed its name to the General Electric Company Ltd. Its trade mark was a horseshoe magnet with a few turns of wire on it, and 'Magnet'

Fig. 15.3 *Selection of early electric heaters in the South Eastern Electricity Board's Milne Museum, Tonbridge.*
Three radiant fires using cylindrical carbon-filament lamps can be seen on the floor at the back. On the left is an early storage heater. Most of the others show various designs with nichrome wire in fireclay supports

appliances were widely advertised. Several British appliance manufacturers began as branches of American firms, including Hoover, Thomson-Houston and Westinghouse. The latter two merged in 1928 as Associated Electrical Industries Ltd. Fig. 15.4 shows electrical appliances from a catalogue of 1906.

In about 1912, A.F. Berry of the British Electric Transformer Company opened 'The Tricity House' in Oxford Street, London. It was a restaurant serving seven hundred meals a day, and also a showroom for electric cooking equipment. Berry lectured on the merits of electric cooking, arguing that one ton of coal delivered to the power station could do as much cooking, by electricity, as ten tons of coal delivered to the house (Fig. 15.5). Furthermore, said Berry, meat shrank much less in an electric cooker than when cooked by a coal fire. A joint of meat weighing eight pounds and cooked by electricity would provide as much meat as a ten pound joint cooked by coal.[4]

The Hotpoint Electrical Appliance Co. Ltd., was founded by an American, Earl H. Richardson, who had been manufacturing electric irons in California before setting up a British company. The name 'Hotpoint' originated when one of his irons developed a fault which led to overheating in the middle of the sole. His wife told him that what housewives really needed was an iron that was hottest at the point.[5] Belling and Company, makers of electric heaters and cookers, were founded at

Edmonton in North London in 1912. C.R. Belling began his experiments in a garden shed with a 10 kW rapid water heater, but the supply system at that time could not accept loads of that size being switched on and off rapidly.

A great advance in electric heating was provided by the development in 1906 of nichrome, an alloy of nickel and chromium. It has the important property that it will not oxidise even when red hot. Most electric fires ever since have used elements of nichrome wire wound on fireclay rods or wound into spirals and stretched in grooves in fireclay supports (Fig. 15.3).

Fig. 15.4   *Merryweather's catalogue, 1906*

The Swedish firm Electrolux exported refrigerators to Britain from 1919 and began manufacturing refrigerators and vacuum cleaners in Britain in 1927. Very few refrigerators were sold in Britain before 1950, but over half the households had one by 1970, and by 1980 more domestic electricity was used for refrigeration (including freezers) than was used for lighting.

The City of London Electric Lighting Company charged 4d (nearly 2p) per kWh for electricity for cooking and heating, which was half their charge for lighting. Elsewhere the difference was often greater. In 1916 the *Electrician* published a detailed survey of all the prices charged by the 564 supply undertakings then operating in the United Kindom. Most quoted at least two domestic tariffs ('lighting' and 'other') and many had three ('lighting', 'power', and 'heating and cooking'). A few also had a 'vehicle charging' tariff, which was always the cheapest. Typical lighting charges were from 3d to 7d per kWh; typical heating and cooking charges were from 1d to 3d. The power charges were usually intermediate.[6]

" TRICITY " OUTFIT IN AN OLD WORLD VILLAGE COTTAGE.

Fig. 15.5 *Photographs from the published version of A.F. Berry's lecture on electric cooking, just before the First World War*

In London suburbs in the 1930s there were usually two tariffs, about 4d for lighting and about 1d for heating, but there were still wide variations between areas.[7] Although virtually all houses built in the London suburbs had electric lighting installed, it was rare for there to be more than three 'power points' rated at 5 or 15 A for electric fires, and a similar number of sockets rated 2 A for a radio or table lamp. In the more expensive houses, however, electric panel fires began to be built in as standard fittings in an least one bedroom.

Despite the efforts of appliance manufacturers and supply undertakers, electric cooking was not adopted quickly (Fig. 15.7). In 1938, when two-thirds of British homes had electricity, only 18% of the electric homes had an electric cooker. By 1970 when nearly every home had electricity the proportion with electric cookers had risen to 40%.

Fig. 15.6   Carron 10 kW electric cooker, in daily use from 1900 to 1975.
(Photo: Electricity Council).

The policy of offering appliances for hire continued until the electricity supply industry was nationalised in 1948. Since then hiring has largely ceased, though Electricity Boards offer hire purchase and other 'easy terms' to encourage customers to buy.

Until the 1950s gas cookers had several advantages over electric cookers, which undoubtedly account for the popularity of gas. The electric cookers had solid iron hotplates, heated by a nichrome element, and were slow to heat up and slow in responding to the controls. Because heat transfer to the saucepan or frying pan was mainly by conduction rather than by radiation, the pans had to have heavy bases which had been machined flat. The switch usually provided only 'high', 'medium' and 'low' settings. The element was in two parts which were connected in parallel

Fig. 15.7 *Electric cooker displayed in Croydon Corporation Electricity Showroom, about 1930*
(From a photograph in the South Eastern Electricity Board's Milne Museum, Tonbridge)

for 'high' and in series for 'low' (in series each element was on half voltage and therefore gave one quarter of the heat it would have given on 'high'). For medium only one part of the element was used. Gas, on the other hand, could heat a light-weight pan and the flame would respond almost instantly to adjustment of the infinitely variable gas tap.

An electric cooker with a rapid-boiling ring in place of one of the usual hotplates was on the market in the 1930s. The ring operated on 12 V from a transformer in the bottom of the cooker, and consisted of a spiral of solid conducting material. The modern radiant ring has a mains voltage element within an earthed metal tube packed with mineral insulation. They operate at temperatures up to dull red heat and pans are heated mainly by radiation. Such rings were first introduced in the late 1950s and rapidly replaced the iron hotplate. Another important change in electric cookers, introduced along with the radiant ring, was the infinitely variable control. Some have used semiconductor switching devices, though most have simple, mechanical switch contacts which are opened and closed every few seconds by a thermal timer. Turning the control alters the on/off ratio of the contacts. In the late 1970s the ceramic hob appeared. Individual hotplates or rings with all their attendant cleaning problems are replaced by a smooth, flat sheet of ceramic material which conducts heat readily through its thickness. Conventional heating elements are clamped to the underside of the ceramic sheet, but the cook sees only a smooth surface with marks to indicate where pans should be stood.

The microwave oven for reheating cooked food was first demonstrated at a catering exhibition in London in 1959. In such ovens an oscillator operating at either 896 MHz or 2450 MHz produces a field which acts directly on the molecules of the food or other material to be heated. This has the result that the food is heated uniformly throughout. In conventional cooking heat is applied to the surface of the food and reaches the inside, slowly, by conduction. Microwave ovens were quickly adopted by the catering trade, since they permitted food to be taken from the deep freeze and made ready to eat in a few minutes. They are only slowly coming into domestic use. Concern about their safety has probably discouraged many potential users, and all microwave ovens are provided with interlocks to ensure that the oven cannot be 'on' unless the door is properly closed.

In 1933 the Electrical Development Association organised a vigorous campaign to promote electric water heating. The EDA had been founded in 1919 by the Institution of Electrical Engineers, Electricity Supply Undertakings (both company and local authority), manufacturers and electrical contractors. Its object was co-operative 'publicity and propaganda work' (the EDA's own expression) on behalf of the industry. They published a large number of advertising and information leaflets mainly aimed at the general public, though some aimed at retailers to encourage them to sell more electrical goods. A booklet issued to the trade in 1933 points out that a house with an electric water heater will use an extra 1000 kWh per year, and it gives a table of the number of water heaters in one town:[8]

1925	nil
1927	465
1928	716
1930	1282
1931	1666
1932	1944

Alas, the town is not named, nor is its size indicated. It was probably a fairly large town, for even in 1948 only 16% of households connected to the electricity supply had water heaters.

The proportion of electric households having an electric iron has always been high, reaching 90% about 1950.

The 'off-peak' period was initially the hours of daylight, but by the 1930s the pattern of demand had changed and the time of reduced demand was during the night. Supply undertakings began to offer 'off-peak' tariffs for loads which were connected through a time switch so that they were only 'on' at night. An obvious application was water heating, though few houses had a hot water tank that was lagged adequately to keep hot all day.

Storage heaters, for room heating, were introduced on a small scale in the 1930s. They contained a mass of brick or similar material in a well-lagged casing. The heater was 'charged' at night and gave out heat during the day. In practice the lagging was inadequate, and only a few storage heaters were used. In the 1960s the Electricity Council conducted research into storage heater design and conducted a vigorous advertising campaign to sell off-peak electricity. In 1962 0·1% of British households had storage heaters. By 1976 the figure reached a peak of 9·1%, though it has since fallen slightly. In addition a smaller number of households have adopted underfloor electric heating, which is also in effect a storage heater system.

It is convenient to mention here a few appliances which do not really come under the heading 'electric heating' but which were, nevertheless, promoted by the supply industry. One such popular domestic electrical is the vacuum cleaner. (Fig. 15.8). In 1938 27% of electric households had one, and the proportion reached 75% in 1961. The vacuum cleaner was invented in 1901 by a young civil engineer, H. Cecil Booth. Booth set up his Vacuum Cleaner Company in London. This had a horse-drawn van, containing the suction equipment driven by a petrol engine, which was brought to the customer's house. Long hoses were taken into the house to clean carpets and furniture. By 1905 Booth had made portable vacuum cleaners which could be operated manually or driven by an electric motor.

Washing machines were rare in Britain before the Second World War, but in half the households by 1963. Most of the 1930s washing machines had no heater but consisted of a tub and agitator driven by an electric motor mounted underneath. Usually a wringer was fixed over the tub and a gear-box directed the drive from the motor either to the agitator or to the wringer. A few spin driers, which spin washed clothes at high speed to remove much of the water by centrifugal action, were made before the Second World War, but they only became common after 1960. By 1970 most households had a spin drier (often combined with a washing machine as a 'twin-tub'), and the wringer to squeeze out the water had virtually disappeared. Automatic washing machines which take in and heat the water, agitate the clothes, and then rinse with fresh water according to a programme, were first available about 1960.

15.3 Industrial electric heating

The iron and steel industry is one of the largest users of electrical energy. Electric arc steelmaking was pioneered by Siemens in Germany around 1880, but

Fig. 15.8   Early vacuum cleaners and other appliances in the South Eastern Electricity Board's Milne Museum, Tonbridge

the process was little used until the First World War. It then became important because of the demand for steel for munitions, and quickly became a major user of electricity. The Bessemer converter, in which the majority of high grade steel had previously been produced, could only accept a limited quantity of scrap, and the introduction of electric arc furnaces made it possible to re-cycle the steel scrap from the munitions factories. There was a large increase in the use of electricity by the steel industry during the 1960s: consumption doubled in just over ten years.

In the electric arc furnace an arc is struck between the steel being melted and a carbon electrode above it. Other methods of direct electric heating in the steel industry include induction heating and direct resistance heating. Both processes create heat within the metal itself. The heating is more uniform than heating in a furnace, and far less heat is wasted. The greater efficiency of direct electric heating compared with heating by coal, oil or gas, can offset the greater cost of electricity, and because there is less waste heat the working conditions are much better.

Apart from actually melting metal, electric heating is used for heating billets prior to rolling or forging. The choice between induction heating and direct resistance heating depends partly on the shape of the billet: in general, long thin billets are best heated by resistance heating and more compact ones by induction. An important advantage of electric billet heating is that very little scale forms on the billet. The waste of material from scaling when a billet is heated in a fuel-fired furnace can be 5%. Electric heating reduces this to below 0·5%.[9]

Steam is used in many industries, often produced by a coal-fired boiler. Electric heating to produce steam was introduced in Switzerland in 1916 because there was a shortage of fuel but an abundance of water power which could be used to generate electricity. They used 'electrode' boilers, which operate by passing a current through the water. Because the resistivity of water varies considerably, depending on the impurities present, some mechanism has to be provided which varies the area of electrode in the water. However, electrode boilers are cheap to instal and maintain since they require no boiler house, flue or fuel store, and they can be placed close to the point where the steam is required.

Electrode boilers were little used in Britain until the 1930s, but were then encouraged when cheap off-peak tariffs became available and they are now widely used in industry.[10]

The expression 'infra-red heating' as applied to industrial processes was first used in the USA in the 1930s when the heat radiated from ordinary light bulbs was used to cure synthetic resin enamels on car bodies. Both carbon filament and tungsten filament lamps were used, and they were run at a lower temperature than when used as light sources. This prolonged the life of the lamps and shifted the lamp output towards the infra-red part of the spectrum. During the Second World War infra-red heating was introduced into Britain and used in the aircraft industry. The chief industrial use is for curing paints and resins and for drying thin materials such as paper. It is also used for heating plastic sheets, and sometimes metals, prior to pressing into shapes.

In the late 1950s infra-red heaters were developed using a nichrome spiral in a silica tube. Apart from industrial applications such heaters have become popular for use in bathrooms and in some public buildings where heating is required only for short periods of time. The fact that the radiation passes through the air and heats directly the person or object it strikes makes infra-red heating economic in situations where most other forms of heating would wastefully heat the air and surrounding building. In industry it is sometimes possible to select the temperature of the heater so that the radiation is at a wavelength that is selectively absorbed by the substance to be heated. For example water has an absorption peak at a wavelength of 3 μm, and infra-red radiation of that wavelength is particularly good for drying.[11]

15.4   References

The statistics quoted in this section are derived partly from Corley, T.A.B.: *Domestic Electrical Appliances*, Jonathan Cape, 1966, and partly from Electricity Council information literature

1   Crompton, R.E.B.: 'Central station lighting: transformers v. accumulators', *J. Soc. Tel. Eng.*, 1888, 17, pp. 349-387
2   Crompton & Co.: *Catalogue of domestic equipment*, 1900
3   Papers and press cuttings in the Crompton Collection in the Science Museum Library (Archive Section) mostly from unidentified sources
4   Berry, F.A.: *Electric cooking and heating*, British Electric Transformer Co. Ltd., Hayes, Middlesex, 1917
5   Corley, T.A.B.: *op. cit.*, pp.30 and 151
6   Supplement to *Electrician*, 18 February 1916
7   Jackson, Alan A.: *Semi-detached London*, George Allen & Unwin, 1973
8   Most of the publications of the Electrical Development Association are in the Electricity Council Archives. The water heating booklet mentioned is EDA leaflet no. 1087, of 1933
9   Edwards, A.W., and Laws, W.R.: *Rapid heating of billets by electricity*, Electricity Council, 1965
    Parkins, Alan: 'Rapid electroheating of steel billets for forging and rolling', *Electr. Rev.*, 5 December 1975, pp. 739-742
10   Edwards, J.C.: *Steam and how to raise it electrically*, Electricity Council, 1961
    Dobie, W.C.: 'Producing process steam and hot water economically', *Electr. Rev.*, 5 December 1975, pp. 748-750
11   Sharples, J.T.: *Infra-red heating*, Electricity Council, 1969

Chapter 16

Motors and traction

16.1 D.C. motors: generators in reverse

The fact that a motor and a generator are the same machines used in different ways may have been recognised by Watkins in 1835. It was certainly appreciated by Antonio Pacinotti who in 1863 introduced a machine which could be used either as a generator or a motor. Pacinotti's machine used the ring-wound armature, subsequently adopted and improved by Gramme and known ever since as the 'Gramme ring'.[1]

The Siemens brothers were also finding that their generators could be used as motors, and in 1872 Werner Siemens wrote to his brother Carl that the 'small rotating machine runs just as well as a motor as it does as a generator . . . I believe that with this we can soon tackle the problem of electric carriages.'[2] The flourishing firm of Siemens & Halske set out to find customers requiring electric power transmission.

The first public exhibition of the electrical transmission of power by means of a generator and a motor was probably at the Vienna Exhibition of 1873. The Gramme company showed two identical machines five hundred metres apart, one used as a generator and the other as a motor driving a pump.[3] Similar demonstrations were given at the Philadelphia Exhibition in the USA and at the Loan Collection of Scientific Apparatus in London, both in 1876.

By 1874 Gramme had electrically driven machines in his Paris factory, though he used a single lighting generator as a motor to turn a line shaft, not individual motors for each machine.[4] Some exhibitors at the Paris Exhibition of 1881 demonstrated motors. Marcel Deprez showed five motor-driven sewing machines, lathes, a drilling machine, and a printing press. Siemens exhibited a lift in the building (Fig. 16.1) and a tram running along the Champs Elysées (Fig. 16.2).[5]

One of the first Gramme machines to be sold as a motor was used to drive a conveyor for unloading beet at a sugar factory in Sermaize, France, belonging to Messrs Chrétien & Felix. This installation was so successful that in May 1879 the company experimented with the use of Gramme motors for ploughing a rectangular

Fig. 16.1   *Siemens' electric lift at the Paris Electrical Exhibition, 1881*
(From the photograph in the IEE Archives)

field nearby. The plough was pulled across the field by cables between two wagons, each carrying a winding drum and a motor. The general arrangement was similar to that used for steam ploughing.[6]

The first company to exploit the commercial potential of electrical motors on a large scale seems to have been the German branch of Siemens. In July 1877 Werner Siemens wrote to his brother Carl in Russia about a visit from the Head of the Prussian State Mines, Krug von Nidda, who wanted to drive drills electrically and also to refine copper electrolytically.

Thus, a new field has opened up for our business. Krug von Nidda already has a double current generator as good as ordered which transmits 5 horsepower. He thinks his compressed air drills receive only 25% usable power and 75% is lost . . . There are innumerable uses for power transmission without shafting.

Between 1877 and 1882 Siemens conducted an extensive research programme on the application of electricity to mining.[7]

Fig. 16.2 *Siemens' electric tram running along the Champs Elysées to the Paris Electrical Exhibition, 1881*
(From a photograph in the IEE Archives)

At the Berlin Exhibition of 1879 he showed an electric locomotive which could haul eighteen passengers in three cars around a circular track 300 m long. A stationary generating plant supplied the locomotive through a third rail, and the running rails provided the return path. The motor of the locomotive was a Siemens' machine, identical to his lighting generators, mounted on a four-wheeled truck. More than 100 000 passengers were carried while the exhibition was open. In 1880 a similar railway was demonstrated at an exhibition in Vienna, though in that case the two running rails were also the supply conductors.

The first permanent electric railway open to the public was built by Siemens & Halske at Lichterfelde in 1881. It ran for about 3 km. Each carriage had a motor under the floor connected to the wheels through a belt drive.[8]

The first electric railway in the United Kingdom ran between Portrush and Bushmills in Ireland. The reason for using electric traction there was that ample water power was available. A hydroelectric generating station was built at Portrush where there was a waterfall on the River Bush. Two 50-horsepower water turbines drove Siemens generators to provide the supply, which was transmitted to the trains through a third rail system, using the running rails for the return circuit. The conductor rail was an iron bar mounted alongside the track. The collectors were steel springs projecting from the side of the train. In some places, such as where roads crossed the line, there was a gap in the conductor rail and the driver had to ensure that he approached the gap with sufficient momentum to cross it. At passing loops one train had to travel a considerable distance without current, but the lines were arranged so that the train in question was going downhill while the train going uphill at the passing loop received a continuous supply.[9]

The first electric railway in England was built on the sea-front at Brighton in 1883 by Magnus Volk, and this railway still runs for the benefit of holidaymakers (Fig. 16.3).

Fig. 16.3   Volk's railway on the sea-front at Brighton in 1883

In 1888 Sprague built the first really practical electric tramway system, in Richmond, Virginia. Forty cars using an overhead conductor system ran in nearly 20 km of streets. To overcome opposition in some places to the use of overhead wires, Sprague equipped some of his tramcars with storage batteries. His most important contribution to electric traction, however, was his development in 1895 of the multiple unit control system. This made it possible to have separate motors distributed along the length of a train but all controlled from a single master controller in the driver's cab. Multiple unit trains required motors that could be built into the bogies. Sprague introduced the construction in which one end of the motor is pivoted on one axle while the other is supported on springs, and the armature is geared to the axle (Fig. 16.4).[10]

Electric trams were introduced more slowly in Britain, partly because the horse tramways already established were nearing the date when they could be purchased compulsorily by the local authorities. New tramways after about 1890 were virtually all electric.

London's first underground railways, the lines now known as the Circle, District, and Metropolitan, were opened from 1863 using steam locomotives. This was possible because the tunnels were 'cut and cover' constructions only just below the

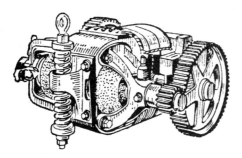

Fig. 16.4 *Sprague's traction motor, designed to fit in a railway bogie*

surface and there were frequent ventilation holes. The first deep 'tube' was the City and South London Railway, running initially from Stockwell to the City about twenty metres below the surface (Fig. 16.5). It opened to the public on 18 December 1890, although the Princess of Wales 'inaugurated' it the previous month. Tube railways are only possible with electric traction and the 'Tuppeny Tube', as it became known since the fare was always twopence, was a great success. It is now part of London Transport's Northern Line. The line was supplied from a single generating station at Stockwell, containing three 300-horsepower reciprocating steam engines and generators which, as mentioned in Chapter 11, were a source of nuisance to the neighbours.

The rolling stock initially consisted of 14 locomotives each of which could haul three carriages with 34 passengers in each. At busy times the service interval between trains was just under four minutes, and five million passengers were carried in the first year. Each locomotive was driven by two series wound d.c. motors whose armatures were built on the wheel axles. The locomotives were reversed by reversing the armature connections: the field polarity was never reversed. Current was collected by an iron shoe from a third rail on glass insulators between the running rails, and the running rails formed the return path. The trains ran at 40 km/h and the combined power of the motors in one locomotive was 75 kW. The supply to the trains was at 500 V.

The Central London Railway (now the Central Line) was also electrified using 500 V d.c. The locomotives had two four-wheel bogies with a motor on each axle. Series parallel control was used: on starting the four motors were all in series across supply; for running at full speed all four motors were in parallel. Starting resistances were also used, and the controller had sixteen steps. Multiple unit control was introduced on the Central London Railway in 1902, and quickly adopted on most of the London Underground.

The first main line railway to adopt electric traction was the Baltimore & Ohio Railroad, in 1892. They were building an extension which included a tunnel where steam traction was undesirable. Electric traction was used through the tunnel, and was so satisfactory that contemporary observers thought it would be extended rapidly. In practice electric traction grew quite slowly.[11]

There were different opinions about the ideal supply voltages for electric trains, and whether the supply should be a.c. or d.c. In 1920 a Railway Advisory Committee recommended the general adoption of 1500 V d.c., which was in use on some main line railways. The Ministry of Transport accepted this and it remained the standard for some years.

Most of the early electric railways and tramways used direct current, because the series-wound d.c. motor has ideal characteristics (especially a high starting torque). For main lines, however, a.c. transmission was preferable, but no a.c. motors were really suitable for the purpose. The introduction of the mercury arc rectifier in 1928 was a landmark in the development of railway electrification. It permitted a.c. supply to the train with rectifiers on the train feeding d.c. motors. Most of British Rail now uses this system with 25 kV overhead wires. The Southern Region of British Rail uses third-rail d.c. at about 700 V. Other countries have used a variety of systems including a.c. at $16\frac{2}{3}$ and 25 Hz and also three-phase a.c. The three-phase systems, in Switzerland and Italy, had one phase earthed to the track and two overhead wires for the other two phases. Since about 1960 the development of germanium and silicon power rectifiers to replace the mercury arc has simplified the rectifying equipment on board trains.

Fig. 16.5 *Drawing of the City and South London Railway, soon after its opening in 1890*

Although d.c. motors and generators are basically the same machines, some design differences appeared quite early. The main visible difference was that motors, which are often used in dirty surroundings, were more enclosed than generators, used in clean engine rooms.

One of the best known British firms was Laurence, Scott and Co. of Norwich. In the 1890s they were supplying totally enclosed motors for factories. These machines naturally ran quite hot, and were said to be good for warming the workers' tea. The founder of the firm, William Hardman Scott (died 1938) began his electrical career with the Hammond Electric Light Company and went to Norwich in 1883 to instal lighting plant for Colman's, the mustard manufacturers. Scott thought he could make better machines than Hammond's, and with financial backing from Colman's he set up a partnership with a Mr. Paris.

Scott and Paris soon linked up with Reginald Laurence, a mechanical and civil engineer with money to invest, and the firm became Laurence, Scott and Paris. By 1900 the name of Paris had disappeared from the Company's literature: it was Laurence, Scott & Co. Ltd. They specialised in d.c. machines, especially for use on ships. A point of interest to historians is that detailed records and drawings of thousands of machines from about 1900 onwards have been preserved. In 1927 Laurence, Scott amalgamated with Electromotors Ltd of Manchester, which made a.c. machines.[12]

In the United States there were fifteen manufacturers of electric motors by 1887, and more than 10 000 machines had been produced. The most important manufacturer was the C & C Company, which began in 1886. The initials stood for C.G.Curtis and F.B.Crocker, and because they were closely associated also with S.S.Wheeler the firm was also known as the Curtis, Crocker, Wheeler Company. Their first motors were used for driving sewing machines, some being wound for operation from a 6 V battery, others for use on the 100 V mains.

Another American pioneer of electric motors was Frank Julian Sprague (1857-1934), who trained as an engineer in the US Naval Academy. He left the Navy in 1884 to organise his own electrical engineering company. One of his first products was an electric hoist for use on building sites, from which he moved on to electric passenger lifts. His major contributions, however, were in the field of electric traction.[13]

D.C. motors are mainly used in traction, and in other applications where speed control is all important, such as lifts and rolling-mill drives. Such machines may be rated at several thousand horsepower, but the principles of the motor are no different from the smaller machines. Control systems are discussed below, but it is appropriate to consider first the early a.c. motors.

16.2 Induction and synchronous motors

The earliest motors operated on direct current but the advantages of a.c. for transmission and distribution encouraged engineers to develop an a.c. motor. The first practical a.c. motors were the induction and synchronous motors developed by Tesla in 1888.[14]

Nicola Tesla (1856-1943) was born in Smiljan, then in Austria-Hungary but now in Yugoslavia. After studying at Graz and Prague he took up electrical engineering. He made his first invention, a telephone repeater, in 1881 while working at a telephone exchange in Budapest. In 1884 he emigrated to the USA, and became an American citizen. He worked for Edison for a few years, and then joined Westinghouse. Edison was a 'd.c.' man, firmly opposed to a.c.; Westinghouse was the leading American exponent of a.c. systems. In changing employers Tesla was stating his own views on the future direction of electrical engineering.

The fact that a pivoted permanent magnet or a pivoted piece of magnetic material will follow a rotating magnetic field had been discovered by Arago in 1824 (see Chapter 2). Arago produced his rotating magnetic field by rotating a permanent magnet. Tesla's great achievement was to produce a rotating magnetic field from alternating currents flowing in fixed coils. His first motor had two field coils energised by two alternating currents whose waveforms were 90° out of step (Fig. 16.6). Tesla showed that the resultant magnetic field of the two coils was roughly constant in strength but rotating in direction. He went on to show that three coils spaced 120° apart and supplied with three-phase a.c. would also produce a rotating field.

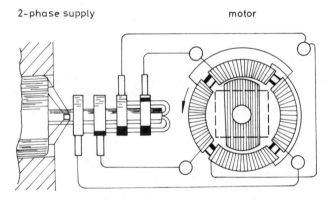

Fig. 16.6 *The principle of Tesla's two-phase induction motor*
A generator (on the left in the drawing) has two sets of coils on its armature, and gives currents 90° out of step. These currents pass through coils arranged at right angles in the motor and produce two oscillating magnetic fields. The resultant of the two magnetic fields is a field of constant strength whose direction rotates. The rotor of the motor follows the rotating magnetic field.

Tesla announced his discovery of how to make a rotating magnetic field at a meeting of the American Institute of Electrical Engineers on 16 May 1888.[15] By 1889 Tesla had obtained ten United States patents, covering two-phase and three-phase induction and synchronous motors, a two-phase four-wire power distribution system, and also split-phase starting of a single-phase motor (see Fig. 16.7).

The synchronous motor corresponds to Arago's experiment with the pivoted permanent magnet. The rotor turns at exactly the same speed as the rotating field although the rotor poles will lag behind the apparent poles of the rotating field by an angle dependent upon the load. The synchronous motor therefore runs at a

Fig. 16.7 *Experimental induction motor made by Tesla about 1888, and given by him to the Science Museum*

constant speed, determined by the frequency of the supply and the number of pairs of poles of the rotating field, and has to be run nearly up to 'synchronous' speed by some other means when first switched on. The induction motor corresponds to Arago's experiment with a disc that is not a permanent magnet. The rotating field induces currents in the disc which react with the field to produce rotations. Since the induction motor depends on relative motion between the rotor and the rotating magnetic field in order to induce currents in the rotor, it follows that the rotor speed must always be a little less than the synchronous speed.

Other people were working on the problem of creating a rotating magnetic field, including Michael Osipowitsch von Dolivo-Dobrowolsky (1862-1919), a Russian by birth who spent much of his life in Germany.[16] Another was the Italian Galileo Ferraris, who published his work just before Tesla. Both Tesla and Ferraris described a motor in their initial publications, but Tesla had developed his ideas much further. When challenged on the ground that Ferraris had anticipated him, Tesla's main patents were upheld in the United States' Courts.[17]

Tesla also showed that an induction or synchronous motor could be run from a single-phase supply if part of the field winding were connected through a capacitor or inductor, to produce a second phase. Once started the motor will run satisfactorily on the single-phase supply, and in many motors the starting capacitor or inductor is taken out of the current automatically once the motor is running.

The Westinghouse Company bought Tesla's patents, and from 1892 they began to promote polyphase a.c. distribution systems. They adopted the three-phase 60-Hz which remains the American standard to this day. European engineers preferred lower frequencies, usually 25 or 50 Hz. An experimental single-phase induction motor made in London about 1891 by W. Langdon Davies is shown in Fig. 16.8. By

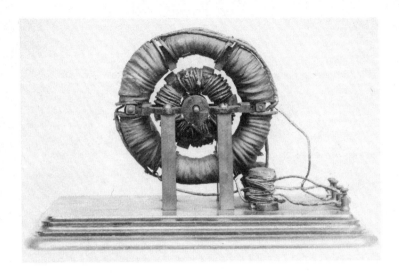

Fig. 16.8   *Experimental single-phase induction motor by W.Langdon Davies about 1891*

1895 Westinghouse had a range of induction motors in production. For motors up to five horsepower the rotor was a 'squirrel cage' construction as used in modern induction motors. The electrical part of the squirrel cage rotor consists of two end rings linked by straight bars. (Squirrels were sometimes kept as pets in the late ninteenth and early twentieth centuries. Presumably the rotor reminded people of the squirrel's cage or, more likely, of the treadmill sometimes placed in pets' cages to amuse the animal — or its owner). For higher rated motors Westinghouse preferred to have the rotor wound with wire. A resistance was connected in series to reduce the starting current, and then cut out automatically when the motor had run up to speed.

The American General Electric Company decided to make a.c. machines, and in 1896 they were offering induction motors up to 150 horsepower.[18]

Most of the world's electric motors today are induction motors. They can only be used where a constant speed is required, but for many industrial applications that is perfectly acceptable. Because it needs no brushgear the induction motor gives reliable service over long periods of time with little or no maintenance. Most domestic washing machines use an induction motor with capacitor start; for other domestic appliances the lighter universal motor is preferred.

An important type of induction motor in low power applications is the 'shaded-pole' motor. As already mentioned, a single-phase induction motor requires a second phase to be introduced for starting, although the machine will then run without it. In the shaded pole motor there is only one field winding but part of the iron pole face is 'shaded' by surrounding it with a thick copper ring. The alternating magnetic flux induces a current in the ring which has the effect of retarding the flux in that part of the pole. In effect the shaded pole motor is a two-phase machine, with the flux in the shaded part of the pole lagging behind the main flux. The resultant flux has a rotating component, so the motor will start. The simple construction of shaded pole motors makes them very reliable. They need no maintenance and are widely used for low power applications where efficiency is not important, such as fans.

A motor which is easy to confuse with the induction motor is the repulsion motor. The confusion is compounded by the existence of mixed action motors which start as repulsion motors but run as induction motors.

The repulsion motor is largely due to J.A.Fleming and Elihu Thomson. Fleming, who was professor of electrical engineering at University College London, is best known for his work in connection with radio. Fleming made a study of the forces between conductors carrying alternating currents. In 1884 he showed that a copper ring suspended within a coil carrying an alternating current tends to twist so as to be edge on to the magnetic field. This is the basis of the repulsion motor. Fleming presented a survey of much of his work in a Discourse at the Royal Institution on 6 March 1891.[19] He demonstrated the 'jumping ring' experiment in which a copper or aluminium ring dropped over a vertical iron bar is thrown off when alternating current is passed through a coil on a bar. By cutting the ring and inserting a lamp he showed that a current flowed around the ring and that the repulsion of the ring was due to this current. Fleming then described the action of induction motors and of repulsion motors, which were being developed from his ideas by Elihu Thomson in America. For the latter he had demonstration equipment sent him by Thomson.

Thomson's repulsion motors had a single field winding and a wound multi-coil armature with a commutator. Two brushes which are connected together act on opposite sides of the commutator and short-circuit the armature coil which is positioned across the magnetic field. There is then a turning force on that coil, and the armature rotates.

More complex repulsion motors were soon made. Some had two sets of field windings placed at right angles and connected in series. The motor could be reversed by reversing the connections of one winding. Some had two sets of short-circuited brushes, or one set of short-circuited brushes and another set connected to the secondary of a transformer whose primary is in series with the field winding. The advantage of this complicated arrangement was that the machine had a very good power factor.

Repulsion motors are mainly single-phase, though can be made for polyphase supplies. They give a good starting torque and have been used extensively for electric traction with speed control obtained by shifting the brushes. In practice large repulsion motors require more maintenance of the brushgear than d.c. machines, and for lower power applications the capacitor-start induction motor or the 'universal' motor is preferred.

16.3   Control of motors

In many applications motors are 'controlled' simply by switching them on or off.

If the starting current is heavy then they may first be connected in series with a resistance, which is then cut out.

Where several d.c. motors are working together, as in an electric train, it is usual to adopt 'series-parallel' operation. If there are two motors and a resistance then the controller will connect the machines to the supply in the following sequence:

motors in series plus resistance
motors in series only
motors in parallel plus resistance
motors in parallel

The principle can be extended if three or more motors are involved. When two motors are connected in series to the supply each is effectively on half voltage, and the starting current is reduced accordingly.

In applications where a single high-power motor is required and where good control is needed over a wide range of speeds, then a d.c. motor and the Ward Leonard control system are normally used. H. Ward Leonard (1861-1915) was an American electrical engineer with a special interest in lifts. He appreciated that the ideal way of controlling a d.c. motor is to control its armature current. In the Ward Leonard system, developed about 1890, a first motor is used to drive a generator and the output of that generator supplies the armature of the motor being controlled. The generator output is regulated by controlling its field current, which is, of course, very much smaller than the motor current. Numerous 'feedback' systems have been devised in which the generator field current is controlled by a device which monitors any deviation of the motor's actual speed from its desired speed.[20]

Ward Leonard controls are often used for winding motors in mines and for rolling mill drives, both applications where fine control is vital. They have also been used in quite low power applications: the Science Museum possesses a Ward Leonard set rated at only a few hundred watts which was used to pull 'silver paper' through a colouring bath.

The induction motor is essentially a constant speed machine, and for many applications that is perfectly acceptable. The pole amplitude modulated, or 'p.a.m.', motor devised by Professor G.H.Rawcliffe in 1957 is an induction or synchronous motor which can run at either of two predetermined speeds. The field windings are so arranged that by changing the connections of a few coils the number of poles can be changed. The p.a.m. motor therefore retains the reliability and robustness of the conventional induction motor while being able to work at two alternative speeds.

Another approach to the ideal of an induction or synchronous motor whose speed can be varied is to provide a variable frequency supply. With power semiconductor devices this becomes feasible. Transistors capable of controlling a few amperes became available in the mid 1950s, and by about 1960 thyristors (then known as silicon controlled rectifiers) were available which could control some tens of amperes. Thyristors are used in inverter circuits to generate variable frequency alternating current for supplying induction motors. They are also used in 'chopper' circuits which switch a circuit on and off rapidly to vary its effective voltage. The speed of d.c. motors may be controlled in this way, and similar systems may be used for controlling lights or anything that responds to a variable voltage. All such control systems are easily regulated by reliable and compact electronic circuits responsive to speed or any other chosen parameter.

16.4 'Universal motors'

The motors described so far have been distinctly 'a.c.' or 'd.c.' machines. The speed of the d.c. motors can be regulated over a wide range. The a.c. induction and synchronous motors are essentially fixed speed machines, and though the speed of a.c. repulsion motors can be varied these machines have other disadvantages.

If a d.c. motor is operated on a.c. the polarities of the field and armature will be reversed at the end of each half cycle, and since the polarities of both are reversed the direction of the motor is unchanged. If the field iron is laminated, as well as the armature iron, so that the alternating field current does not produce eddy currents in the iron, then the machine will run satisfactorily on alternating or direct current supplies. This is the 'Universal' or 'a.c./d.c.' commutator motor which is widely used in applications such as electric drills, food mixers and vacuum cleaners. An advantage of the universal motor is that it can run much faster than an induction motor, and therefore have greater power for a given weight of machine.

The idea of combining an electric motor in a single housing with the device it drives dates from about 1895, when an electric drill was made in Germany. The Wolf electric drill, first produced in Britain in 1914, was available in d.c., single-phase a.c., and three-phase a.c. versions. The power was 370 W and the weight over 10·7 kg, giving a power-to-weight ratio of 35 W/kg.

In 1925 Wolf introduced a drill using a universal motor. It had a power of 100 W and weighed 2½ kg, a power-to-weight ratio of 40 W/kg. That machine ran at 1500 r.p.m. on load. By 1940 Wolf was making a drill running at 2750 r.p.m. which had a power of 175 W and weighed 1·1 kg, a power-to-weight ratio of 150 W/kg. This machine was designed for use by women in factories during the war. After the war similar drills were sold widely to home handymen.

Almost every domestic motorised appliance, except the washing machine, uses a universal motor built in to the appliance. Such equipment has become smaller and lighter as the motor has been developed.

Until the 1950s the electrical safety of appliances usually depended on enclosing the 'live' parts in metal casing, which was earthed. Although appliances with an earthed casing are intrinsically safe if properly used, there were felt to be two hazards in practice. It was possible to connect the earth wire incorrectly, so that the casing was made 'live', and there was no means of ensuring that the earth connection was sound. A few appliances were fitted with a neon lamp connected between the live wire and the casing: the lamp should light if, but only if, the appliance were properly connected and the user should always check that the neon lamp was alight before using the appliance.

An alternative approach to the question of safety was the concept of 'double insulation'. In this the motor is first insulated in its own housing and then a second insulating barrier is placed around it. 'Double insulation' was first adopted in Germany in 1953, and accepted five years later in Britain. In both cases the system was first used in an electric drill, but it quickly became the norm for most domestic appliances.

16.5   Electric road vehicles

At the turn of the century electric road vehicles were more common than petrol driven ones. There were two distinct types: the self-contained machine supplied from batteries, and the trolley vehicle drawing current from overhead wires. By 1900 there were several hundred electric taxi-cabs in New York and other American cities, and a smaller number in London.[21] Electric vehicles gave a service which was considered quite adequate until the First World War, when petrol driven vehicles were able to travel further without refuelling. Supply undertakings often charged a special low tariff for vehicle charging, to encourage electric cars.

The first trolley bus service began in 1901 in Bielethal in Germany, though an experimental trolley vehicle ran in America in 1889 (Fig. 16.9). The first trolley buses in Britain appeared in 1911, and in London in 1931. Just after the Second World War the number of trolley buses in the world reached a peak of about 6000, but the number has been falling ever since. The main problem seems to be the inflexibility of the trolley bus, which can only go where its overhead wires have been installed.

Trams and trolley buses have now virtually disappeared from the streets of Britain, though in many European cities a fleet of modern, smooth-running trams, sometimes articulated, provides excellent public transport.

There are still many electric vehicles in daily use, usually small delivery vehicles such as milk floats. Most electric vehicles use heavy lead-acid or nickel-iron batteries. Much research effort is currently going into the quest for a better battery. A possible contender is the sodium-sulphur battery, which offers a much lighter weight battery than conventional types. It suffers from the fundamental disadvantage that it operates at about 300°C, and hot sodium could be dangerous in an accident. Petrol is also dangerous. It may well be that further research will find a battery, using the sodium-sulphur system or some other chemical combination, that will again give electric vehicles an advantage over petrol ones.

The Electricity Council currently sponsor electric vehicle research, but the supply industry has been doing so at least since 1913. In June of that year the Incorporated Municipal Electrical Association decided to appoint a Committee 'to consider and adopt such measures as may be found desirable to further the use of the electric battery vehicle'. Its chairman was R.A.Chattock, the city electrical engineer of Birmingham and President of the IMEA, and the membership included representatives of company-owned electric supply undertakings and manufacturers of storage batteries and electric vehicles. Its aims included such practical measures as establishing standards for charging arrangements, as well as persuading supply undertakings to provide charging stations. It also sought to induce insurance companies to quote favourable rates for insuring electric vehicles 'in consequence of the reduced fire risk'.

The *Electrical Review* welcomed the formation of the new committee, and urged that supply undertakings should provide charging facilities without waiting for the introduction of vehicles. 'It is obvious that if the supply authorities wait for the vehicles to arrive before they provide the charging stations, the latter will never be required'.[22]

DIBBLE'S ELECTRICALLY PROPELLED CARRIAGE.

Fig. 16.9 *Dibble's electrically propelled carriage, from Scientific American of July 1889*

Regrettably, the Committee seems not to have achieved anything. It was eventually absorbed in the Electrical Development Association and, as already stated, the research continues.[23]

16.6 Linear motors

Linear motors are often described as 'ordinary' rotary motors which have been split along their length and unrolled. It follows that there are as many kinds of linear motors as there are of rotary ones. Linear motors may be a.c. or d.c. commutator machines, or they may be induction or synchronous machines, to name just a few possibilities.

The description of a linear motor as an unrolled rotary machine was used by Wheatstone in 1841, and he made a linear motor whose stator survives (Fig. 16.10). It was a d.c. commutator machine based on his electromagnetic engine designs, and with the aid of a suitable armature and brushes it can be demonstrated. The stator is a solid piece of iron 30 x 12 x 2·5 cm with slots cut in it which hold twelve coils. The ends of the coils are brought out to thirteen brass contact pieces. With carbon brushes held on the contact pieces by hand, and moved lengthwise, a piece of steel tube can be made to roll along the stator. Since it requires a current of at least 30 A to achieve this it is doubtful whether Wheatstone himself ever saw his machine work.[24]

The idea of a linear motor has re-appeared several times since Wheatstone's attempt in 1841. His friend William Henry Fox Talbot (1800-1877) patented a machine in 1852 which had a row of horseshoe magnets energised successively to make an iron cylinder roll across them. Fox Talbot is best known as the inventor of photography, but his interests included electricity and he worked with Wheatstone on a number of electrical experiments. His most unusual motor was his 'electrolytic gas engine', designed in consultation with Wheatstone, and made by W.T.Henley. The machine had a piston moving in a vertical cylinder and coupled to a crank. Two electrodes at the bottom of the cylinder permitted a liquid such as acidulated water to be electrolysed, and then a heater wire higher in the cylinder ignited the resulting gases. Two such machines were made, though there is no evidence that Fox Talbot ever succeeded in making either of them run.[25]

Wheatstone's and Talbot's linear motors were 'magnetic' machines, in which an iron armature was attracted to a solenoid. The first demonstration of linear motion produced by electromagnetic induction was Fleming's 'jumping ring', mentioned above in connection with the repulsion motor. Fleming, however, seems not to have seen the linear possibilities of his device, and he immediately turned his attention to the problem of obtaining rotary motion by means of the mechanism he had demonstrated. That was in the early 1890s. Ten years later Fleming had turned his attention to radio, and the leading figure in a.c. machine development was the French engineer P. Boucherot. He is best known for his invention of the double-cage rotary induction motor, but he was also fascinated with the prospect of a practical linear motor. He lamented that in any motor the magnetic pull between rotor and stator was of the order of 20 to 40 times the tangential thrust, yet only the latter was usable. He designed reciprocating machines which used the attractive force, converting it to rotary motion by a ratchet device, though none achieved lasting acceptance. The idea was revived in the early 1970s when Hesmondhalgh, Tipping and Sarson of the University of Manchester published a detailed study of such a ratchet motor. They noted that a major design problem in making a ratchet motor was moving the armature away from the electromagnet after it had been attracted towards it. Their solution was to build a machine with the armature held

Fig. 16.10 *Stator of Wheatstone's linear motor of 1841*

by springy members to give a mechanical oscillatory system. The armature was set into vibration at the natural frequency of the system, and drove the output shaft through a ratchet.[26]

The Norwegian Professor Kristian Birkeland obtained a series of patents between 1901 and 1903 for a d.c. linear motor used as a silent gun.[27] It was not successful, probably because being a 'magnetic' rather than an 'electromagnetic' machine it was incapable of being scaled up satisfactorily, but the idea of a linear motor gun has recurred periodically. The Russian engineer N. Japolsky worked on linear motors in Russia around 1930. According to Professor Laithwaite[28] Japolsky believed that the first sputniks were launched with the aid of linear motors. In later life Japolsky lectured on electromagnetic theory at King's College London, where the present writer knew him; his work with linear motors was not mentioned.

Japolsky was probably wrong in thinking that linear motors were used to launch sputniks, but it is interesting that he thought it possible. During the Second World War the American Westinghouse Company built a linear motor aircraft launcher, called the 'Electropult'. The aircraft sat on a trolley with windings underneath, and the fixed part also had windings. The Electropult produced 10 000 horsepower and could accelerate a 4½ tonne aircraft to 180 km/hr in 4·2 seconds.

The two potential applications of linear motors that have attracted most attention throughout the twentieth century are driving shuttles in a loom and railways. Emile Bachelet set up a company to work on both applications in 1914. Eric Laithwaite was first interested in the linear motor for use in a loom, and turned later to its use in transport. The loom requires a means of projecting the shuttle at high speed across the width of the cloth. Conventionally, this is done by striking the shuttle very hard and catching it on the other side, a process which wastes energy and requires considerable mechanical strength in the shuttle. Despite many attempts linear motors have still not taken over in this application.

The story of linear motors in transport is one of promising but unsuccessful ventures. Readers interested in this theme are referred to Professor Laithwaite's book in this series.

One application where linear motors have achieved considerable success is in pumping and stirring liquid metals. In the 1930s pumped liquid metal was used for heat transfer in special circumstances, where the high thermal capacity of the metal was useful, and where direct contact with the liquid metal was to be avoided. The liquid metal itself acts as the moving part of the motor, and all that is necessary to pump the metal is to fix a wound stator onto the wall of the container holding the metal. Such devices are now commonly used for stirring aluminium and other metals in furnaces, and for assisting the transfer of metal to moulds for casting. A very specialised application is for pumping the liquid sodium coolant in some nuclear reactors. Reliable operation and complete isolation of the metal being pumped are vital requirements, and the linear motor provides these.

16.7   The versatility of the electric motor

Electric motors are now made in a vast range of types and sizes. Large motors drive trains, ships, or rolling mills and may be rated at thousands of horsepower. At the other end of the scale an electric analogue watch has a motor that takes so little power it can run for a couple of years on a tiny cell. These small machines are electromagnetic engines, the direct descendants of the machines described in Chapter 4 and of the motors in the ABC telegraph receivers.

Nowhere is the versatility of electric power seen better than in the electric motor. The reader is invited to count the electric motors in his or her home, then to add the electric motors in the factories that made the things in the home, and the electric motors in the vehicles that bring things to the home. Most gas or oil-fired central heating systems have an electric pump, and every petrol or diesel vehicle has an electric motor for starting, and usually others for operating windscreen wipers, blowers etc.

Electric motors make the world go round!

16.8 References

1 For Watkins see Chapter 4; for Pacinnotti and Gramme see Chapter 6
2 Sigfrid von Weiher and Herbert Goetzler, *The Siemens Company: its historical role in the progress of electrical engineering,* Berlin and Munich, 1977, p.36
3 *Engineering,* 1879, **28,** p.417. E Hospitalier, *Les Principales Applications de l'Electricité,* 1881, Paris, p.340
4 *Engineering,* 1879, **28,** p.417
5 *Engineering,* 1881, **32,** p.567, and photographs of the exhibition in the IEE Archives
6 *Telegraphic Journal and Electrical Review,* 1879, **7,** p.299.
   Wormell, R.: *Electricity in the service of man,* revised edition, Cassell & Co, 1896, p.647 (There are several editions of this book from 1888 onwards).
   La Lumière Electrique, 1879, **1,** p.47
   Hospitalier, *op. cit.,* p.341 and planche IV
7 Von Weiher and Goetzler, *op. cit.,* p.36, quoting (in English) letter dated 13/14 July 1877 from Werner Siemens to Carl Siemens, and also quoting Sigfrid von Weiher, 'Werner Siemens and the introduction of electrical engineering into mining and metallurgy', *Bergfreiheit,* 1953, **18,** pp.388-392
8 Wormell, *op. cit.,* p.684: fuller details are given in earlier editions
9 *Electrician,* 1883, **10,** p.545
10 MacLaren, Malcolm: *The rise of the electrical industry during the nineteenth century,* Princetown University Press, Princetown, New Jersey, 1943, p.94
11 MacLaren, *op. cit.,* p.104
12 *Scott built a Dynamo,* Laurence, Scott & Electromotors Ltd., Norwich, 1965
13 MacLaren, *op. cit.,* pp.92-96
14 There are several biographies of Tesla, including *Nikola Tesla: Life and work of a genius,* a collection of essays on aspects of Tesla, published by the Yugoslav Society for the Promotion of Scientific Knowledge, Belgrade, 1976
15 *Journal of the American Institute of Electrical Engineers,* 1888, **5,** p.308
16 *Electrician,* 1891, **27,** p.389
17 *Electrician,* 1895, **36,** p.281
18 MacLaren, *op. cit.,* pp.98-99
19 Fleming, J.A.: 'Electro-magnetic repulsion', *Electrician,* 13 March 1891, pp.567-571, and 20 March 1891, pp.601-604
20 Ward Leonard, H.: 'Speed regulation of electric motors', *Transactions of the American Institute of Electrical Engineers,* 1896, **13,** pp.377-386, and obituary of Ward Leonard, *Electrician,* 30 April 1915, p.115
21 Montague, Lord, of Beaulieu, 'Electric vehicles' in *Oxford History of Technology, VII,* 1978, pp.731-733
22 *Electr. Rev.,* **73,** 26 September 1913, Report of IMEA with detailed aims of the electric vehicles committee, p.496; editorial comment, p.481-482
23 *Electr. Rev.,* 18 December 1925, p.987
24 Bowers, Brian: *Sir Charles Wheatstone,* HMSO and Science Museum, 1975, pp.82-84
25 Bowers, Brian: 'The electrolytic gas engine', *Proc. Weekend Meeting on History of Electrical Engineering,* IEE, 1977, paper 2
26 Hesmondhalgh, D.E., Tipping, D., and Sarson, A.: 'High-torque low-speed motor using magnetic attraction to produce rotation', *Proc. IEE,* 1973, **120,** pp.61-66
27 Norwegian patents 11201 and 11342 of 1901; 11228 of 1902; 13035 and 13052 of 1903

Chapter 17

The electrical world

17.1 Summary

Power cuts are rare events nowadays, but when they occur we realise how dependent we have become on public electricity supply.

I woke this morning to an electric alarm clock, turned on the light and the radio and plugged in the kettle. I could hear the whine of the milkman's electric van and the hum of the tumble drier, controlled by a time-switch to use off-peak electricity while we sleep. After an electric shave and breakfast from the toaster and percolator, I set off by electric train to the office, leaving my wife to switch on the dishwasher and type the remaining pages of this book on her electric typewriter. So the day continues: nor does sunset curtail our activities. We may watch television or play records. We may use brighter lighting to read or sew or help the children work out control circuits for their model railway. Finally we retire to bed – electrically heated if necessary.

Victorian electrical engineers saw that electricity was destined to bring great changes, but they would be amazed if they could see how great those changes have been. Electricity was seen as an agent to bring light and heat without smoke and dirt. An article in the *Electrician* in 1879 deplored the inefficiency of small steam engines and the pollution caused by the thousands of such engines in use. The writer, who was probably Oliver Lodge, thought the power of flowing rivers could be harnessed in preference to coal.

> We degrade a large population to the most loathsome labour in the pits, we consume and poison the air, and we load it with such quantities of smoke that the sun is barely visible . . .
>
> Large steam engines will for a long time remain; small ones will be superseded by electromagnetic engines driven by electricity, supplied either by water power or by a large central coke-burning steam engine. Gas will supersede the crude burning of coal for cooking and warming purposes, and a smokeless millenium will set in.[1]

Hydroelectric power has not in fact achieved much in England, though he was right in predicting that large central power stations would make more efficient use of fuel. In the year ended 31 March 1980 77% of the Central Electricity Generating Board's production came from coal fired plant, 11·7% from oil, 10·4% from nuclear power, and 0·9% from natural gas. Seven years earlier, before the massive increase in oil prices that began in 1974, the total fuel consumption was 4% less and the proportions were coal 62·5%, oil 26·5%, nuclear 8·6%, natural gas 2·4%.

Lodge could not have foreseen the introduction of nuclear power into the commercial electricity supply system; Berkeley and Bradwell nuclear stations in England were commissioned in 1962. At 31 March 1980 nine of the CEGB's 132 stations in service were nuclear. At the same date ten power stations were under construction, of which three were nuclear. Future power station policy (how many power stations should be built and what fuel they should use) is a matter of current public debate.

A pioneer who saw clearly what electricity might achieve was R.E.B. Crompton, who said in 1890,

England in future, instead of being spoilt by densely populated industrial centres, might be covered with cottages extending for miles over the present almost uninhabited rural districts, so that the population would be more evenly spread over the kingdom. The factory hands, instead of having to work under the shafting in factories, should be able by the electrical transmission of power to carry on industrial pursuits in their own cottage homes. That is the future which lies before electrical engineers if they have the pluck and energy to force their views upon the public to a sufficient extent.[2]

When the Electrical Development Association was set up after the First World War only 12% of households in Great Britain had electricity supply. The theme that recurs through the Development Association's advertising between the wars is that electricity is the answer to the servant problem. About 1930 the Association compiled a lecture 'Transforming the homes of Britain' intended 'for delivery before groups of the general public'. It said:

The domestic servant is vanishing, soon she may have vanished altogether. Before the war the proportion of women who were willing to enter domestic service was steadily declining, and since the war the households which possess one or more really competent servants are in the minority. Most people, apart from those able to pay high wages and offer easy conditions, have to make the best of more or less incompetent help for a limited number of hours each day, and thousands of women are confronted with the almost impossible task of simultaneously doing the work, or most of the work, in their homes and bringing up a family.[3]

The proportion of households with electricity rose steadily until in 1938 65% were connected. After the supply industry was nationalised in 1948 a major effort was made to extend the supply to every household. Crompton's vision of factories giving way to cottage industry has not come true, but electric power in the home has become a vital part of everyday life.

In the 1980s a large number of women go out to work as well as run their homes and bring up a family. Whether or not this is desirable is outside the scope of this book; undoubtedly the variety of 'electric servants' in the home has helped to make this possible.

In the years 1979/80 and 1980/81 the demand for electricity in England and Wales has fallen slightly compared with the previous year. For a century the industry has believed that demand would increase year by year for evermore. The current decline may be a short-term aberration or the supply industry may have reached the ultimate maximum demand. Whether or not there will be a demand for *more* electricity in future, there will be a need for even finer control of the electricity we use. Electronic and microelectronic devices are being used increasingly in machines in factories and in appliances in the home, to make them ever more versatile and sensitive to the needs of the consumer. Thus a rolling mill drive in a steelworks can sense the thickness of the rolled plate, or a tumble drier can decide for itself when the washing is adequately dried.

The electrical engineer has wrought great changes in the last hundred years, and his ingenuity has not yet reached its limit.

17.2   References

1   Quoted and attributed to Lodge in Jolly, W.P.: *Sir Oliver Lodge*, Constable, 1974, pp. 43-44
2   Crompton, R.E.B.: *Reminiscences,* Constable, 1928, pp.226-7
3   EDA leaflet no. 1119 in the Electricity Council Archives

Index